KNOWLEDGE - BASED INTELLIGENT TECHNIQUES in CHARACTER RECOGNITION

Edited by
Lakhmi C. Jain
Beatrice Lazzerini

CRC Press
Taylor & Francis Group
Boca Raton London New York

CRC Press is an imprint of the
Taylor & Francis Group, an **informa** business

CRC Press
Taylor & Francis Group
6000 Broken Sound Parkway NW, Suite 300
Boca Raton, FL 33487-2742

First issued in paperback 2019

© 1999 by Taylor & Francis Group, LLC
CRC Press is an imprint of Taylor & Francis Group, an Informa business

No claim to original U.S. Government works

ISBN-13: 978-0-8493-9807-0 (hbk)
ISBN-13: 978-0-367-39978-8 (pbk)
Library of Congress Card Number 99-12428

Library of Congress Cataloging-in-Publication Data

Knowledge-based intelligent techniques in character recognition /
    [edited by] L.C. Jain, B. Lazzerini.
        p.   cm. -- (The CRC Press international series on
computational intelligence)
Includes bibliographical references and index.
ISBN 0-8493-9807-X (alk. paper)
    1. Optical character recognition devices. 2. Writing-
-Identification--Data processing. 3. Expert systems. 4. Soft
computing. I. Jain, L. C. II. Lazzerini, Beatrice, 1953–
III. Series.
TA1640.K56 1999
006.4′24—dc21   99-12428
                                                                       CIP

Visit the Taylor & Francis Web site at
http://www.taylorandfrancis.com

and the CRC Press Web site at
http://www.crcpress.com

The CRC Press
International Series on Computational Intelligence

Series Editor
L.C. Jain, Ph.D., M.E., B.E., (Hons), Fellow I.E. (Australia)

L.C. Jain, R.P. Johnson, Y. Takefuji, and L.A. Zadeh
Knowledge-Based Intelligent Techniques in Industry

L.C. Jain and C.W. de Silva
Intelligent Adaptive Control: Industrial Applications

L.C. Jain and N.M. Martin
Fusion of Neural Networks, Fuzzy Systems, and Genetic Algorithms: Industrial Applications

H.N. Teodorescu, A. Kandel, and L.C. Jain
Fuzzy and Neuro-Fuzzy Systems in Medicine

C.L. Karr and L.M. Freeman
Industrial Applications of Genetic Algorithms

L.C. Jain and Beatrice Lazzerini
Knowledge-Based Intelligent Techniques in Character Recognition

L.C. Jain and V. Vemuri
Industrial Applications of Neural Networks

H.N. Teodorescu, A. Kandel, and L.C. Jain
Soft Computing in Human-Related Science

B. Lazzerini, D. Dumitrescu, L.C. Jain, and A. Dumitrescu
Evolutionary Computing and Applications

B. Lazzerini, D. Dumitrescu, and L.C. Jain
Fuzzy Sets and Their Application to Clustering and Training

L.C. Jain, U. Halici, I. Hayashi, S.B. Lee, and S. Tsutsui
Intelligent Biometric Techniques in Fingerprint and Face Recognition: Practical Applications

Z. Chen
Computational Intelligence for Decision Support

L.C. Jain
Evolution of Engineering and Information System and Their Applications

H.N. Teodorescu and A. Kandel
Dynamic Fuzzy Systems and Chaos Applications

L. Medsker and L.C. Jain
Recurrent Neural Networks: Design and Applications

PREFACE

The main aim of this book is to present results of research on intelligent character recognition techniques. There is tremendous worldwide interest in the development and application of personal identification techniques. Personal identification techniques are useful in many situations including security, bank check processing and postal code recognition. Examples of personal identification techniques are character recognition, fingerprint recognition, face recognition and iris recognition. Most character recognition techniques involve data acquisition, preprocessing, feature extraction and classification.

The main objective of automating the character recognition process has been to make the automated system more 'human like.' The primary techniques employed to achieve 'human like' behavior have been fuzzy logic, neural networks and evolutionary computing.

This book has eleven chapters. The first chapter, by Jain and Lazzerini, introduces principles of signature verification for on-line and off-line systems. Some considerations regarding performance evaluation of handwriting recognition systems are presented. The second chapter, by Shouno, Fukushima and Okada, shows that neocognitrons trained by unsupervised learning can recognize real-world handwritten digits provided by a large database. The third chapter, by Satoh, Kuroiwa, Aso and Miyake, presents a novel rotation-invariant neocognitron. This rotation-invariant recognition is implemented by extending the neocognitron, which can recognize translated, scaled and/or distorted patterns from those used in training.

The fourth chapter, by Baldwin, Martin and Stylianidis, discusses the integration of soft computing to develop a system for recognizing handwritten numerals. The fifth chapter, by Sorbello, Gioiello and Vitabile, presents algorithms for recognizing handwritten characters using a multilayer perceptron neural network. The sixth chapter, by Liu and Fung, presents a technique for automatically identifying signature features and verifying signatures with greater certainty. The use of fuzzy-genetic algorithms overcomes traditional problems in feature classification and selection, providing fuzzy templates for the identification of the smallest subset of features.

The seventh chapter, by Banarse and Duller, presents the application of a self-organizing neural network to handwritten character recognition. The eighth chapter, by Lazzerini, Reyneri, Gregoretti and Mariani, presents an Italian bank check processing system. This system consists of several processing modules including those for data acquisition, image preprocessing, character center detection, character recognition, courtesy amount recognition and legal amount recognition. The ninth chapter, by El-Yacoubi, Sabourin, Gilloux and Suen, presents an off-line handwritten word recognition system using hidden Markov models. Chapter ten, by Malaviya, Ivancic, Balasubramaniam and Peters, presents off-line handwriting recognition systems using context-dependent fuzzy rules. The last chapter, by Brugge, Nijhuis, Spaanenburg and Stevens, presents a license-plate recognition system for automated traffic monitoring and law enforcement on public roads.

This book will be useful to researchers, practicing engineers and students who wish to develop a successful character recognition system.

We would like to express our sincere thanks to Berend Jan van der Zwaag, Ashlesha Jain, Ajita Jain and Sandhya Jain for their assistance in the preparation of the manuscript. We are grateful to the authors for their contributions, and thanks are due to Dawn Mesa, Mimi Williams, Lourdes Franco and Suzanne Lassandro for their editorial assistance.

L.C. Jain, Australia
B. Lazzerini, Italy

CONTENTS

Chapter 1:

An Introduction to Handwritten Character and Word Recognition

AN INTRODUCTION TO HANDWRITTEN CHARACTER AND WORD RECOGNITION

L.C. Jain
Knowledge-Based Intelligent Engineering Systems Centre
University of South Australia
Adelaide, Mawson Lakes, SA 5095, Australia

B. Lazzerini
Dipartimento di Ingegneria della Informazione
Università degli Studi di Pisa
Via Diotisalvi 2, 56126 Pisa, Italy

There is a worldwide interest in the development of handwritten character and word recognition systems. These systems are used in many situations such as recognition of postcodes, interpretation of amount and verification of signature on bank checks, and law enforcement on public roads. The tremendous advances in the computational intelligence techniques have provided new tools for the development of intelligent character recognition systems. This chapter introduces principles of handwriting recognition for on-line and off-line systems. Some considerations about performance evaluation of a handwriting recognition system are discussed.

1   Introduction

In the last few years many academic institutions and industrial companies have been involved in the field of handwriting recognition. The automatic recognition of handwritten text can be extremely useful in many applications where it is necessary to process large volumes of

0-8493-9807-X/99/$0.00+$.50
© 1999 by CRC Press LLC

handwritten data, such as recognition of addresses and postcodes on envelopes, interpretation of amounts on bank checks, document analysis, and verification of signatures [1]. Substantial progress has been recently achieved, but the recognition of handwritten text cannot yet approach human performance. The major difficulties descend from the variability of someone's calligraphy over time, the similarity of some characters with each other, and the infinite variety of character shapes and writing styles produced by different writers. Furthermore, the possible low quality of the text image, the unavoidable presence of background noise and various kinds of distortions (such as poorly written, degraded, or overlapping characters) can make the recognition process even more difficult. Therefore, handwriting recognition is still an open and interesting area for research and novel ideas.

Handwriting recognition systems acquire data by means of either off-line devices (such as scanners and cameras) or on-line devices (such as graphic tablets). In off-line recognition systems, the input image is converted into a bit pattern. Then, specific preprocessing algorithms prepare the acquired data for subsequent processing by eliminating noise and errors caused by the acquisition process. In contrast, on-line recognition systems use dynamic writing information, namely, the number, order, and direction of the strokes to recognize each character as it is being written. Typically, the two-dimensional coordinates of the pen position on the writing surface are sampled at fixed time intervals.

The field of automatic recognition of handwritten isolated *characters* is the longest established branch of handwriting recognition. A character is traditionally identified by the set of its fundamental components, or *features*. A great variety of types of features have been used, including topological and geometrical, directional, mathematical, and structural features. Several conventional and computational intelligence techniques have been applied to character recognition such as statistical analysis, structural and syntactic analysis, pattern matching, neural networks, fuzzy systems and genetic algorithms.

A more difficult field is the recognition of handwritten *words*, especially fully cursive (i.e., with a fully connected main body) words. Two approaches to word recognition are mainly used; the global approach and the analytical approach. The *global* (or *holistic*) approach recognizes the whole word. Each word is represented in terms of global features (such as the sequence and position of vertical strokes, ascenders, descenders, and loops) and the representation of the word is compared with models, describing the ideal shape of the possible words, stored in a reference lexicon [2]. One or more models may be built for each word to take into account the wide variability of cursive script. The global approach is usually adopted in applications with a reduced size (not exceeding a few hundred words) lexicon (for example, in check processing).

The *analytical* approach divides a word into smaller units, typically characters or groups of characters (*segmentation* process). The isolated units are then individually recognized and fit within a possible word.

Of course, the recognition accuracy of a handwriting recognition system is strongly dependent on the segmentation correctness; particularly, an erroneous segmentation may lead to incorrect recognition. A wrong segmentation may be produced, for example, by the lack of a clear separation between consecutive characters, or by the fact that the same sequence of strokes can be interpreted (also by human readers) in different ways.

To solve, at least partially, the aforementioned problems, a third approach to word recognition, named *integrated segmentation and recognition* (or *recognition-based* approach), is sometimes adopted. It consists in tightly integrating word segmentation with character recognition, without requiring any preliminary word segmentation.

It is well known that robust and efficient handwriting recognition can only be achieved by the use of *context*; i.e., information external to the characters [3], [4], [5]. In fact, even humans can make mistakes when reading without utilizing context because of the geometric diversity of character shapes produced by different writers. The problem is even worse when reading handwritten text by computer owing to imperfect writing instruments, scanning mechanisms, learning techniques and recognition algorithms.

Using context means exploiting domain-specific information to make the recognition process easier. There are basically three types of knowledge: morphological, pragmatic and linguistic [6]. *Morphological* knowledge refers to the shape of an ideal (i.e., writer-independent) representation of the characters. *Pragmatic* knowledge refers to the spatial arrangement of ideal character representations into words and phrases. *Linguistic* knowledge refers to the specific language used and concerns the lexical, grammatical, and semantic aspects. In particular, syntax and semantics are used to describe how words and phrases in a document are related. Pragmatic and linguistic knowledge can be considered contextual knowledge.

Examples of the use of contextual knowledge are the following. Simple reading mistakes can be detected and corrected by verifying recognized words through a dictionary, the recognition of handwritten ZIP codes can be made easier by verifying the digits with the address information, etc. Contextual knowledge can also be used to ensure proper word identification in the case in which only parts of the handwritten text image have been recognized because of missing characters, noise or distortions.

2   Knowledge-Based Intelligent Techniques

The knowledge-based intelligent techniques [7] include expert systems, neural networks, fuzzy systems and evolutionary computing. These techniques try to mimic the performance of biological systems in a very limited sense. Expert systems solve specific problems in a given domain in the same way as human experts. Artificial neural networks are designed to mimic the biological information processing mechanism. Evolutionary computing is used in implementing many functions of intelligent systems including optimisation. Fuzzy logic provides a basis for representing uncertain and imprecise knowledge. The majority of chapters in this book demonstrate the application of knowledge-based intelligent techniques in character recognition.

3   Off-Line Handwriting Recognition

Most current approaches to off-line handwriting recognition consist of three main phases, namely, *preprocessing*, *segmentation*, and *recognition*. In describing these phases, we will refer to a generic passage of text. Furthermore, being convinced that an automatic reading system should take soft decisions (as the human cortex does), we assume that most modules (like the character and word recognizers) of such a system would output a list of candidates, possibly ordered by decreasing confidence values, instead of just one.

3.1 Preprocessing

The preprocessing phase aims to extract the relevant textual parts and prepares them for segmentation and recognition. (Note that, depending on the adopted method, some of the preprocessing actions described below may be omitted as well as performed in a slightly different order). First of all, the image may be physically rotated and/or shifted for slant and skew correction. To this aim, the average slope of horizontal/vertical strokes in the specific text is estimated using a Hough transformation of the contour or by computing directional histograms of pixel densities. Then the reference base line of each text line is found, which detects the area where the handwritten text is located.

To remove noise from the image, the document image is filtered. Then, it is thresholded to leave every pixel black or white. A thinning algorithm is finally used to reduce the width of lines to 1 pixel while preserving stroke connectivity. Thinning finds the skeleton of a character image to make identification of character features easier. The raw skeletons are then smoothed by removing edge points.

The next crucial point is the removal of printed boxes and horizontal guidelines used to indicate the position of the information fields to be filled in manually by the writer. Examples of such fields are the courtesy (digit) amount, the legal (worded) amount and the customer's signature fields in bank checks. As the guidelines can be overlapped by the handwritten text, some textual pixels could be deleted as well. A connected stroke or line segment might be broken into two or several parts, resulting, for instance, in the letter "b" to be transformed into "h". This kind of ambiguity is solved by matching with words in a dictionary.

Several different approaches to guideline removal have been proposed that limit handwritten text erosion [8], [9]. Typical examples are the morphological erosion and dilation operators [10]. In general, however, simple heuristic criteria can help to distinguish guidelines from text. For example, variations in the average direction of the guideline indicate the presence of pixels belonging to the text.

Horizontal density histograms can also be used to detect the upper, middle, and lower zones of a word. The first and last zones contain, respectively, ascenders and descenders. The second zone contains characters like "a", "c", "e", etc. Word zoning information is useful to determine the type of a word, that is, whether it is made of middle zone letters only or it also includes letters with ascenders and/or descenders. Also, singularities of handwriting (corresponding to key letters) can be detected by using vertical histograms within the upper or lower zone. Similar techniques can be used to estimate the character height to provide scale invariance. Finally, conversion techniques from two-dimensional to one-dimensional data could be adopted to exploit on-line recognition methods.

3.2 Segmentation

The text lines are segmented into sentences, words and characters. Sentences are supposed to be terminated by punctuation marks. Typically, potential cuts into words of a given sentence are generated based on the distances between adjacent characters (of course, a blank space is a clue to word separation). The probability (or confidence) for each potential cut to be a correct sentence segmentation is calculated. Then, one or more possible sentence segmentation options are considered. Finally, the most probable segmentation options give origin to sequences of possible words. Each such word is input to the word recognizer. The word recognizer, exploiting one of the techniques previously introduced, outputs an ordered list of the best n words (candidate words) in the lexicon which correspond to the unknown word. The list is sorted according to word confidence values. Of course, the word recognizer may contain a segmentation process at the character level.

For each sentence segmentation option, a list of candidate sentences is generated from the combination of the corresponding candidate words and verified by means of syntactic/contextual analysis. Sentence confidence values are associated with the resulting correct sentences.

Of course, the confidence value associated with an entity (sentence, word, etc.) should take into account the confidence values associated with its constituent sub-units.

3.3 Recognition

Essentially, this phase aims to recognize characters and words (note that most applications of handwriting recognition involve text consisting of just one or a few words). In what follows, we will be concerned with word recognition (Sections 3.3.1, 3.3.2, 3.3.3) and character recognition (Section 3.3.4). As previously stated, a word recognizer can apply one of three possible strategies, namely, analytical, holistic and recognition-based [11], [12].

3.3.1 Analytical Methods

The analytical approach segments a word into smaller, easier to recognize units. These units can be of different types according to the adopted method: segments, characters, graphemes, pseudo-characters, etc. It is essential that these units be strictly related to characters so that recognition does not depend on a specific vocabulary.

The analytical approach performs an *explicit* (or *external*) segmentation, because words are explicitly divided into characters (or other units) which are recognized separately. Of course, context-based post-processing is usually required to ensure proper word recognition. Potential breaking points (like ligatures between characters) are found, mainly based on heuristic rules derived from visual observation. Ligatures may be detected using techniques that identify the difference between the upper and lower contours of the word (ligatures correspond to small values of such difference) [13], or the vertical projection of pixels (again, small vertical projection values are a clue to the presence of ligatures) [14].

In general, these heuristics may cause the following errors: i) *over-segmentation* (a given character is divided into many segments), and ii) *under-segmentation* (a character is not revealed). These errors may have a particularly frequent occurrence in unconstrained handwriting, in which characters in a word may touch or even overlap. In the first case, a filtering algorithm is required to discard redundant cuts (for example, comparing the inter-cut distance with the average width of characters). In the second case, the missing character can be detected by a search in a dictionary.

It may be noted, however, that some letters, such as "m", "n", "u", etc., are almost always segmented into fragments. This fact should be taken into account when searching for a word in the dictionary (i.e., two consecutive characters "i" should be considered equivalent to "n", and so on).

3.3.2 Holistic Methods

A *holistic* method recognizes words as a whole, without preliminary segmentation. It consists of two steps: i) feature extraction, and ii) comparison of the representation of the unknown word with the writer-independent reference representations stored in the lexicon. Efficient methods based on Dynamic Programming (an optimization technique) have been proposed for whole-word recognition. These methods can be divided into two distinct groups: the first group is based on distance measurements; the other is based on a probabilistic framework [15].

Basically, a distance is used to assess a similarity degree between two strings of symbols representing, respectively, the recognizer output and the word stored in the lexicon. The similarity degree is defined as dependent on the minimum number of elementary transformations necessary to transform the first string into the second, and vice versa. Examples of elementary transformations are the deletion, insertion, or substitution of a symbol.

Most probabilistic methods use Hidden Markov Models (HMM), which are tools for modeling sequential processes in a statistical and generative way. Word images can be considered as sequences of observations, which are features like characters, parts of characters, pixel columns, or overlapping sliding windows. So, HMMs can be used to model the events (such as insertion or deletion of a feature) which generate the typical handwriting variations for word descriptions [16].

3.3.3 Recognition-Based Methods

A *recognition-based* method performs a sort of *implicit* (or *internal*) segmentation. Indeed, each potential segmentation point is validated by an isolated character recognizer so that only meaningful segmentation decisions are taken. The recognizer quantifies the "goodness" of a break. Typically, a sliding variable-sized window is used to scan the word and suggest potential cuts. Of course, these methods avoid introducing segmentation errors at the expense of heavy computation.

3.3.4 Character Recognition

Recognition of handwritten characters is undoubtedly the basic task of most automatic reading systems. Isolated character recognition can be regarded as a classical pattern recognition problem requiring feature selection, feature extraction, learning process and classification. Two different approaches to character and numeral recognition can be adopted: global pattern matching and structural analysis. Features are extracted from the original raw image, the thinned image, or the normalized image. Examples of features of the first and second types are the height and the width of the character, the number and position of loops, concavities, convexities, etc. Examples of features of the third type are the positions of the terminal strokes in the main directions, and histograms of horizontal and vertical projections of black pixels.

In general, cursive handwritten character recognition has to deal with such problems as characters damaged by the guideline removal, characters whose shape was altered by the image transformations, and touching characters. In particular, recognition of touching characters requires a preliminary segmentation process that returns the sub-images corresponding to the individual characters. In fact, the segmentation problem is a central one in handwriting recognition.

To separate a pair of touching or connected (by a ligature) characters, we have to detect the touching point or the ligature. Several segmentation strategies have been proposed [11]. As previously observed, heuristics are quite often used, resulting in efficient but error-prone methods. Vertical histograms, and the upper and lower image profiles, together with heuristics, are also used. Alternatively, recognition-based methods provide multiple segmentation candidates and use recognition to choose the best ones. Each character recognizer should be designed to output a list of possible candidate characters sorted according to character confidence values.

4   On-Line Handwriting Recognition

While off-line recognition deals with two-dimensional images, on-line recognition is performed on a one-dimensional representation of the input. In the on-line case, the temporal or dynamic components of the handwriting can be recorded. This has several advantages: writing order, velocity and acceleration of the pen are available and can be exploited by the recognition process. Also, pen-downs and successive pen-ups can be used for stroke detection.

Most approaches to on-line handwriting recognition fall into two main categories: i) analysis by synthesis, which is based on the modeling of handwriting generation [6], and ii) application of off-line recognition methods to on-line recognition.

5   Some Considerations about Performance Evaluation

When dealing with automatic handwriting recognition, usual performance indexes are the following: recognition rate, accuracy rate, substitution rate, error rate, and rejection rate. The meaning of these concepts depends on the specific application and product. For a system designed to read general documents, the recognition rate is the percentage of correctly recognized characters/words of all the characters/words in the document. On the other hand, in a system for check processing, the check amount must be correctly recognized or rejected. Thus the recognition rate of isolated characters is not appropriate for performance evaluation in this case. Instead, the check recognition rate is used. The check recognition rate is the percentage of correctly recognized checks in a set of test checks.

Also, in an automatic bank check processing system, the error rate (i.e., the percentage of wrong amounts) must be as small as possible, often not greater than 0.01%. Therefore, a reasonably high rejection rate must be allowed, that is, when courtesy and legal amounts are not recognized with a high confidence, or inconsistencies between them are discovered, the check reader must be able to refuse to give an answer. In this case, we can calculate the accuracy rate as the ratio between the correctly recognized amounts and the non-rejected amounts. Of course, recognition rate and accuracy rate coincide when rejection rate equals zero.

As far as substitution rate is concerned, one can observe that a few substitution errors (i.e., a few characters/words substituted for the correct ones) will not affect the recognition of a generic document. On the contrary, the substitution error of a check would cause severe legal and business problems; thus, the check substitution rate must be 0. In order to achieve 0 check substitution rate, it is usually impractical to rely on 0 substitution error at the digit/character level. Instead, a cross validation of legal and courtesy amount fields is usually adopted.

In general, to meet such requirements as reliability and robustness in spite of bad quality data, and extremely low error rates, it is essential to use high-accuracy algorithms for handwriting recognition, exploiting all the contextual knowledge available at different processing levels.

It has been clearly shown that the combination of several different recognizers, independently designed by different methods and using different features, ensures as low as possible substitution rates, significant increase in the recognition rate, and high reliability of the recognition system as the errors made by the recognizers should be mostly uncorrelated [17]. The independent recognizers can be considered as experts performing a classification of the unknown pattern on their own. Typically, in correspondence with each detected pattern, each expert produces a list of "candidate" patterns sorted according to a confidence value. To obtain a global classification, the expert responses are combined using one of several combination methods. Various techniques have been proposed for combining complementary classifiers including neural networks and rank order techniques [18].

Another way to improve recognition performance consists in training the system on the specific handwriting. However, the system loses its flexibility and reduces its own capability of recognizing the handwriting of other writers (unless a dedicated writer-dependent training phase is provided). In this regard, a promising approach is represented by self-learning systems, i.e., systems that are designed to recognize omni-writer documents, but can learn from the text itself to recognize the peculiarities of the specific handwriting [19].

6   Summary

This chapter has presented a basic introduction on character recognition. The following chapters cover a broad spectrum of techniques for recognising characters. These chapters include recognition of handwritten digits by a neocognitron, recognition of rotated patterns using neocognitron, soft computing approach to handwritten numeral recognition, handwritten character recognition using a MLP, signature verification using fuzzy genetic algorithm, generic neural network application to handwritten digit classification, integration of neural networks and contextual analysis for recognizing handwritten amounts on checks, HMM-based word recognition, off-line handwriting recognition with context-dependent fuzzy rules, and license-plate recognition using neural network.

References

[1] Impedovo, S. (Editor) (1994), *Fundamentals in Handwriting Recognition*, Springer-Verlag, Berlin.

[2] Powalka, R.K., Sherkat, N., Whitrow, R.J. (1997), Word Shape Analysis for a Hybrid Recognition System, *Pattern Recognition*, Vol. 30, No. 3, pp. 421-445.

[3] Cohen, E. (1994), Computational Theory for Interpreting Handwritten Text in Constrained Domains, *Artificial Intelligence*, Vol. 67, pp. 1-31.

[4] Crowner, C., Hull J. (1991), A Hierarchical Pattern Matching Parser and its Application to Word Shape recognition, *Proc. Intl. Conf. on Document Analysis and Recognition*, Vol. 1, pp. 323-331.

[5] Moreau, J., Plessis, B., Bougeois, O., Plagnaud, J. (1991), A Postal Check Reading System, *Proc. Intl. Conf. on Document Analysis and Recognition*, Vol. 1, pp. 758-766.

[6] Parizeau, M., Plamondon, R. (1995), A Fuzzy-Syntactic Approach to Allograph Modeling for Cursive Script Recognition, *IEEE Trans. on PAMI*, Vol. 17, pp. 702-712, July.

[7] Jain, L.C. (Editor) (1997), *Soft Computing Techniques in Knowledge-Based Intelligent Engineering Systems*, Springer-Verlag, Heidelberg.

[8] Dimauro, G., Impedovo, S., Pirlo, G., Salzo, A. (1997), Automatic Bankcheck Processing: A New Engineered System, *International Journal of Pattern Recognition and Artificial Intelligence*, Vol. 11, pp. 467-504, June.

[9] Govindaraju, V., Srihari, S.H. (1992), Separating Handwritten Text from Interfering Strokes, *From Pixels to Features III – Frontiers in Handwriting Recognition*, Impedovo, S. and Simon, J.C., Eds, Elsevier, pp. 17-28.

[10] Dimauro, G., Impedovo, S., Pirlo, G., Salzo, A. (1997), Removing Underlines from Handwritten Text: An Experimental Investig-

ation, *Progress in Handwriting Recognition*, Downton, A.C. and Impedovo, S., Eds, World Scientific, Singapore, pp. 497-501.

[11] Casey, R.G., Lecolinet, E. (1995), Strategies in Character Segmentation: A Survey, *Proc. of Intl. Conf. on Document Analysis and Recognition*, pp. 1028-1033.

[12] Lu, Y., Shridhar, M. (1996), Character Segmentation in Handwritten Words – An Overview, *Pattern Recognition*, Vol. 29, No. 1, pp. 77-96.

[13] Fujisawa, H., Nakano, Y., Kurino, K. (1992), Segmentation Methods for Character Recognition: From Segmentation to Document Structure Analysis, *Proc. of the IEEE*, Vol. 80, No. 7, pp. 1079-1091.

[14] Su, H., Zhao, B., Ma, F., Wang, S., Xia, S. (1997), A Fault-Tolerant Chinese Check Recognition System, *International Journal of Pattern Recognition and Artificial Intelligence*, Vol. 11, pp. 571-593, June.

[15] Lecolinet, E., Baret, O. (1994), Cursive Word Recognition: Methods and Strategies, *Fundamentals in Handwriting Recognition*, Impedovo, S. Ed., Springer-Verlag, Berlin, pp. 235-263.

[16] Gilloux, M. (1994), Hidden Markov Models in Handwriting Recognition, *Fundamentals in Handwriting Recognition*, Impedovo, S. Ed., Springer-Verlag, Berlin, pp. 264-288.

[17] Xu, L., Krzyzak, A., Suen, C.Y. (1992), Methods of Combining Multiple Classifiers and their Applications to Handwriting Recognition, *IEEE Trans. on SMC*, Vol. 22, No. 3, pp. 418-435.

[18] Ho, T.K., Hull, J.J., Srihari, S.N. (1994), Decision Combination in Multiple Classifier Systems, *IEEE Trans. on PAMI*, Vol. 16, pp. 66-75.

[19] Lazzerini, B., Marcelloni, F., Reyneri, L.M. (1997), Beatrix: A Self-Learning System for Off-Line Recognition of Handwritten Texts, *Pattern Recognition Letters*, Vol. 18, pp. 583-594.

Chapter 2:

Recognition of Handwritten Digits in the Real World by a Neocognitron

RECOGNITION OF HANDWRITTEN DIGITS IN THE REAL WORLD BY A NEOCOGNITRON

H. Shouno and **K. Fukushima**
Department of Systems and Human Science,
Graduate School of Engineering Science,
Osaka University
1-3 Machikaneyama, Toyonaka, Osaka 560-8531, Japan

M. Okada
Kawato DynamicBrain Project,
Exploratory Research for Advanced Technology (ERATO)
Japan Science and Technology Corporation (JST)
2-2 Hikaridai, Seika, Soraku, Kyoto 619-0288, Japan

The neocognitron is a hierarchical neural network model of the primate visual system. It has an ability to recognize visual patterns by learning. The ability of the neocognitron to recognize patterns is influenced by the selectivity of feature-extracting cells in the networks. This selectivity can be controlled by the threshold of these cells. We use dual thresholds in the learning and recognition phases. Thresholds in the learning phase are set high enough for feature-extracting cells to capture small differences of the features. On the contrary, in the recognition phase, the thresholds are set low to obtain the generalization. Using a large-scale database, we show that the neocognitron trained by unsupervised learning can recognize handwritten digits in the real world. The recognition rate is as high as over 98% with a dual threshold in the learning and recognition phases.

1 Introduction

A number of methods have been developed for off-line character recognition. In this chapter, we discuss a method of recognition using an artificial neural network called "neocognitron". Neocognitron has a hierarchical neural network, that is, a model of a primate visual system [1]. It can acquire an ability to recognize visual patterns by learning.

The purpose of this chapter is to demonstrate that the neocognitron can robustly recognize a large set of patterns encountered in the real world. The neural network's ability to recognize patterns is influenced by the selectivity of feature-extracting cells in the networks. We previously showed, using a simple and small database, that the use of different thresholds in learning and recognition phases produces good recognition ability in the neocognitron [2]. This chapter examines and demonstrates the performance of the neocognitron with dual thresholds using a large database, ETL-1. The ETL-1 database is a handwritten character database that contains varieties of handwritten digits freely written by 1,400 people. We use 200 randomly sampled digit-patterns for each category (2,000 patterns in total) for learning and optimization. In addition, another 300 digit patterns for each category (3,000 patterns in total) are used to test the generalization ability. The result of the recognition rate is over 98 % for novel digit patterns in ETL-1.

Incidentally, LeCun et al. [3] also showed a large back-propagation network trained to recognize digit patterns. The structure of their network is similar to the neocognitron. They use a zip code database provided by the U.S. Postal service, and the misclassification rate of their network was 0.14 % for the training set, and 5.0 % for a novel test set.

2   Neocognitron

Figure 1 shows the structure of the neocognitron used in this experiment. The neocognitron consists of two types of cells. We call them 'S-cells' and 'C-cells' [1]. We call a layer consisting of S-cells 'S-cell layer' and that of C-cells 'C-cell layer'.

In this chapter, we denote the S-cell layer as U_{Sl} and the C-cell layer as U_{Cl}, where l is the hierarchical stage number. The term 'lower stage' means a smaller l, and 'higher stage' means a larger l. Each layer consists of 'cell-planes'.

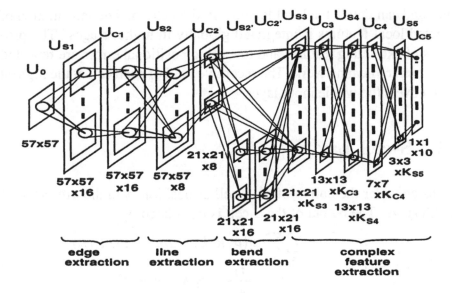

Figure 1. Structure of neocognitron in this experiment. (modified from [6])

A cell-plane has the same type of cells arranged in a two-dimensional array. All the cells in a cell-plane receive input connections that have the same spatial distribution. The positions of the preceding cells from the lead connection shift in parallel with location of the cells in the cell-plane. Therefore, firing of a cell in a cell-plane codes the preferred feature and the position of the feature in the input pattern.

An S-cell models a 'simple cell' in the primary visual cortex [4][5] which works as a feature detector. Input connections to an S-cell are variable and are reinforced by learning. After learning, an S-cell will respond only to a specific feature.

The C-cell is a model of a 'complex cell' in the primary visual cortex. Each C-cell in the same cell-plane receives the input signals from only one S-cell-plane. The C-cell-plane spatially blurs the response of S-cell-plane output. C-cell layers are inserted in the network for tolerating positional errors in the features.

All of the connections of each cell are spatially local in the neocognitron. Therefore, cells in lower stages, such as U_{S1}, have a small receptive field, and can detect the local features in the pattern. The information of detected local features is integrated globally in higher stages. The processes of extracting features and tolerating positional errors are repeated in the hierarchical network. The receptive field of each cell at the highest stage covers the whole input layer.

2.1   S-cell Layer

Let us consider the output of an S-cell at position n in the kth plane of the lth stage. The output of the S-cell is calculated by

$$u_{Sl}(n,k) = r_l \cdot \varphi \left[\frac{1 + \sum_\kappa \sum_\nu a_l(\nu,\kappa,k) \cdot u_{Cl-1}(n+\nu,\kappa)}{1 + \dfrac{r_l}{1+r_l} \cdot b_l(k) \cdot u_{Vl}(n)} - 1 \right], \quad (1)$$

where $a(\nu,\kappa,k)$ denotes the connection weight strength from the κth cell-plane in the preceding C-cell layer(U_{Cl-1}). The numerator in function φ represents the sum of excitatory input and the denominator represents the inhibitory input. Function φ is defined by

$$\varphi(x) = \max[x, 0], \quad (2)$$

which is a half wave rectifying function. Therefore, the final output of S-cells are non-negative. The connections a_l, and b_l are variable connections, and these connections decide the behavior of the S-cell and its cell-plane. The inhibitory input u_{Vl} is

$$u_{Vl}(n) = \sqrt{\sum_\kappa \sum_\nu c_l(\nu) \cdot \{u_{Cl-1}(n+\nu)\}^2}, \quad (3)$$

where $c_l(\nu)$ represents the strength of the fixed excitatory connections, and it is a monotonically decreasing function of $\|\nu\|$.

2.2   C-cell Layer

The output of a C-cell at position n in kth cell-plane of lth stage is described as

$$u_{Cl}(n,k) = \psi \left[\sum_\nu d_l(\nu) \cdot u_{Sl}(n+\nu,k) \right], \quad (4)$$

where d_l is the connection from an S-cell-plane in the same stage. We choose connection $d_l(\nu)$ as a non-negative and monotonically decreasing function of $||\nu||$. Function ψ is a suturation function defined by

$$\psi[x] = \frac{\varphi[x]}{1 + \varphi[x]}, \tag{5}$$

where φ is a half-wave rectifying function defined by Equation (2). A cell in a kth plane receives input signals from only the cells in the preceding kth S-cell-plane, and connections are spatially localized. In other words, a C-cell calculates the local weighted mean of the preceding S-cells' responses and outputs it through the saturation function. The C-cell can robustly respond with tolerance to positional errors in features detected by the preceding S-cells.

In higher stages of the network, such as U_{C3}, U_{C4}, and U_{C5} in this experiment, the spatial density of cells in each cell-plane is decreased. The dimension of the extracted feature is then compressed with small loss of information.

2.3   Detailed Network Architecture of Lower Stages

The initial stage of the network, U_0 layer, is the input layer consisting of a two-dimensional array of receptor cells. The input patterns are presented to this layer. Input patterns are gray valued (8bit) patterns, but almost all the background values are nearly 0. We linearly map these values to $[0, 1]$.

The input connection and threshold of S-cells in U_{S1}, U_{S2}, and $U_{S2'}$ are fixed and predetermined. U_{S1} layer consists of edge extracting S-cells. The input pattern is decomposed into edges of 16 orientations.

The S-cells, in the next stages U_{S2}, extract 8 orientations of line components using the edge information extracted in U_{C1}. They receive signals from a pair of cell-planes in U_{C1} whose preferred orientations differ by π from each other.

$U_{S2'}$ layer, which is in a detouring path between layers U_{S2} and U_{S3}, is a bend extracting layer. In this chapter, 'bend' means an end point of a line and also a point where a curvature drastically changes. U_{S3} layer receives input connections from U_{C2} and $U_{C2'}$. The S-cells in layers higher than U_{S3} have variable input connections, which are determined through learning.

3   Unsupervised Learning of Feature-Extracting Cells

The S-cells in layers U_{S3}, U_{S4}, and U_{S5} have variable input connections which are modified through unsupervised learning with a kind of winner-take-all process [6]. After the learning process, S-cells are able to extract specific features contained in the input pattern.

The learning is performed from lower to higher stages. The training of a stage begins when training of the preceding stages has been completely finished. In this experiment, the method of unsupervised learning is the same as the one used in the conventional neocognitron [1][7]. In the training, 'seed cells', which determine the training feature, are automatically selected with a type of winner-take-all competitive rule.

Figure 2 shows the method of selecting seed cells. Let us consider the situation for reinforcing S-cells in the U_{Sl} layer. Suppose the preceding S-cells layers (U_{S1} through U_{Sl-1}) have already been finished. Then, the response up to the preceding C-cell layer U_{Cl-1} can be calculated. The reinforcement algorithm is indicated as below.

Reinforcement of connections

Require: All the S-cell layers $U_{S1} \sim U_{Sl-1}$ are already reinforced.
Ensure: Reinforce the variable connections a_l and b_l using a training pattern
  1: Input a training pattern.
  2: Calculate responses of the preceding layer U_{Cl-1}.
  3: $K_{Cl-1} \leftarrow$ the number of cell-plane in U_{Cl-1}.
  4: **if** S-cell-plane in U_{Sl} already exists, **then**
  5:    $K_{Sl} \leftarrow$ the number of cell-planes in U_{Sl} layer.

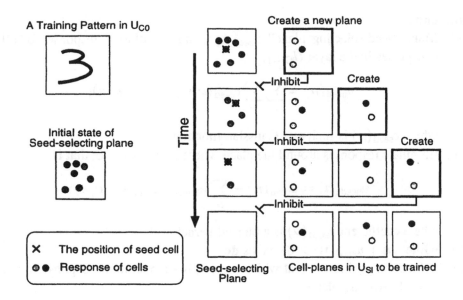

Figure 2. The mechanism of seed cell selection.

6:    Calculate responses of cells in U_{Sl} layer using the current a_l and b_l,
7:    **while** $\{n, k \mid u_{Sl}(n, k) > 0\}$ exists **do**
8:        Search the n^*, and k^* where

$$u_{Sl}(n^*, k^*) \geq u_{Sl}(n, k) \quad \text{for } \forall(n, k).$$

9:        Reinforce the connection a_l and b_l by

$$a_l(\nu, \kappa, k^*) = a_l(\nu, \kappa, k^*) + q \cdot u_{Cl-1}(n^* + \nu, \kappa), \qquad (6)$$

for κ such that $1 \leq \kappa \leq K_{Cl-1}$, and

$$b_l(k^*) = b_l(k^*) + q \cdot \sqrt{\sum_{\kappa} \sum_{\nu} c_l(\nu) \cdot \{u_{Cl-1}(n^* + \nu, \kappa)\}^2}. \quad (7)$$

10:

$$u_{Sl}(n, k) = u_{Sl}(n, k) - u_{Sl}(n, k^*)$$

for k such that $1 \leq k \leq K_{Sl}$ and $k \neq k^*$.
11:        Eliminate responses in k^*th S-cell plane.
12:    **end while**
13:    Calculate responses of cells in U_{Sl} layer using the current a_l and b_l.

14: **end if**

15: Make "seed-selecting plane" u_{seed} by summing blurred responses of all cell-planes in the layer U_{Cl-1}.

$$u_{\text{seed}}(n) = \sum_{\kappa} \sum_{\nu} c_{\text{seed}}(\nu) u_{Cl-1}(n + \nu, \kappa),$$

where $c_{\text{seed}}(\nu)$ is such as a Gaussian function.

16: Eliminate responses around the place where $u_{Sl}(n, k) > 0$; that is,

$$u_{\text{seed}}(n) = u_{\text{seed}}(n) - \sum_{\nu} e_{\text{seed}}(\nu) u_{Sl}(n + \nu, k)$$

where coefficient $e_{\text{seed}}(\nu)$ is a blurred kernel.

17: **while** $\{n \mid u_{\text{seed}}(n) > 0\}$ exists **do**

18:     Find the position n^{**} at which the cell has a maximum response in the seed-selecting plane

$$u_{\text{seed}}(n^{**}) \geq u_{\text{seed}}(n) \text{ for } \forall n.$$

19:     $K_{Sl} \leftarrow K_{Sl} + 1$.

20:     Create a new cell-plane in the U_{Sl} layer.

21:     Prepare a memory for new connections a_l^{new} and b_l^{new}, and clear them.

22:     Reinforce these connections, the a_l^{new} and b_l^{new}, by Equation (6) and (7) under the condition $n^* = n^{**}$.

23:     Eliminate responses of cells around n^{**} in the seed-selecting planes.

24: **end while**

The seed-selecting plane decides the position n^{**}, which indicates the new feature position. This position indicates the center of a training feature. Each input connection to the trained cell is reinforced by the amount proportional to the intensity of the response of the preceding cells. In the process of training, we assume that the other cells in the cell-plane automatically have their input connections reinforced in the same way as the seed cell. In the computer simulation, all the cells in a cell-plane share the same set of connections. Therefore, this condition is automatically satisfied.

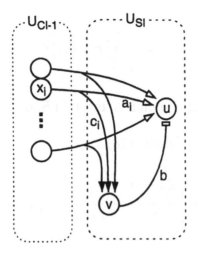

Figure 3. Connections to an S-cell from C-cells in the previous layer.

4   Threshold of S-cells

A feature-extracting cell usually accepts a certain amount of deformation in shape of the feature. The amount of accepted deformation, that is the selectivity of the cell, depends on the threshold of the cell. In this section, we show the relationship between thresholds and selectivity of S-cells.

4.1   Mathematical Notation of an S-cell

In neocognitron, the response of a feature-extracting S-cell can be represented by Equation (1). For the sake of simplicity, we will consider the output of only one S-cell. Figure 3 shows the connections between an S-cell and the preceding C-cells. The responses of C-cells in the preceding stage are represented by a vector x. The output of an S-cell can be acquired by rewriting Equation (1) as

$$u = r\varphi \left[\frac{1 + \sum\limits_{i} a_i x_i}{1 + \dfrac{r}{1+r}bv} - 1 \right], \tag{8}$$

where

$$v = \sqrt{\sum_i c_i \cdot \{x_i\}^2}.$$

(9)

The running index i corresponds to the ν in Equation (1).

We can further rewrite Equation (8)

$$u = \gamma\varphi \left[\frac{\sum\limits_i a_i x_i}{bv} - \frac{r}{1+r} \right] \quad \text{(if } bv > 0\text{)},$$

(10)

where

$$\gamma = \frac{rbv}{1 + \dfrac{r}{1+r}bv}.$$

(11)

Connections a_i and b are determined by learning, and the conventional reinforcement rule for this notation is described as below. If the S-cell becomes a winner in competition for pattern x, the connections a and b are updated by

$$\begin{aligned} \Delta a_i &= q \cdot x_i \\ \Delta b &= q \cdot v \ . \end{aligned}$$

On the contrary, if the S-cell becomes a loser, connections a and b are not updated.

Now let this S-cell be a winner for K patterns $(x^1, x^2, \cdots, x^K)$ in the training pattern set. Connections a_i and b after finishing the learning can be described as

$$\begin{aligned} a_i &= q \sum_k^K c_i \cdot x_i^k \\ b &= q \sum_k^K \sqrt{\sum_i c_i \cdot \{x_i^k\}^2}. \end{aligned}$$

(12)

Substituting Equation (12) into Equation (8), we obtain

$$u = \gamma\varphi \left[\frac{\sum_k \sum_i c_i x_i^k \cdot x_i}{\sum_k \sqrt{\sum_i c_i \{x_i^k\}^2} \sqrt{\sum_i c_i \{x_i\}^2}} - \frac{r}{1+r} \right]. \tag{13}$$

Here we define an (weighted) inner product between two vectors, x and y, by

$$(x, y) = \sum_i c_i x_i y_i.$$

Then, Equation (13) can be rewritten as

$$u = \gamma\varphi \left[\frac{\|X\|}{\sum_k \|x^k\|} \frac{(X, x)}{\|X\| \cdot \|x\|} - \frac{r}{1+r} \right], \tag{14}$$

where

$$X = \sum_k x^k.$$

Here, we define three variables, s, λ, and θ

$$s = \frac{(X, x)}{\|X\| \cdot \|x\|}, \tag{15}$$

$$\lambda = \frac{X}{\sum_k \|x^k\|}, \tag{16}$$

and

$$\theta = \frac{r}{1+r}. \tag{17}$$

Now, Equation (8) can be rewritten as

$$u = \gamma\lambda\varphi \left[s - \frac{\theta}{\lambda} \right]. \tag{18}$$

After enough training, the inhibitory connection b becomes large, and we have $bv \gg 1$. Therefore, Equation (11) can be approximated by

$$\gamma \sim (1 + r).$$

The parameter γ can be regarded as a constant.

4.2 Relationship Between Threshold and Selectivity of an S-cell

We can interpret that the direction cosine s defined by Equation (15) is a measure of similarity between the reference vector and the input vector. The S-cell responds only when the similarity between input vector x and the reference vector X is larger than the θ/λ. Figure 4 illustrates a multi-dimensional vector space, in which the set of connections described by reference vector X and the input pattern described by the input vector x is expressed by an arrow. The cell responds when the test vector x falls into the hatched circular region (tolerance area) surrounding the reference vector X. If the input vector is outside of this area, however, the S-cell does not respond.

The size of the tolerance area is controlled by threshold parameter θ in Equation (17). When the threshold θ of an S-cell is set lower, the tolerance area becomes larger, and the S-cell accepts larger deformation. On the contrary, when the higher threshold value θ is set, the tolerance area becomes smaller, and the S-cell responds only to less deformed test patterns [2].

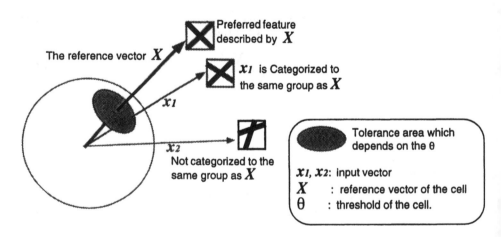

Figure 4. Relationship between threshold of feature-extracting cell and tolerance for deformation

4.3   Calculation of Inhibitory Variable Connection

In the conventional learning method [1][7], λ in Equation(15) is given by Equation (16). Although the value of λ resides in the range $\lambda \leq 1$, it changes depending on the training vectors given to the cell. If λ takes a smaller value, the effective threshold θ/λ becomes larger.

The value of λ depends on the spreading of the training vector falling onto the tolerance area on the sphere surface. If all the training vectors are similar to each other, then $\lambda \sim 1$. When the spreading of falling vectors becomes large, however, the λ becomes small and the effective threshold becomes large. So, the λ is an uncontrollable parameter, which is not desirable. Therefore, we modify the method of reinforcement of the inhibitory variable connections so that $\lambda = 1$ always holds [8] [9]. In the modified learning method, the value of the inhibitory connection b is calculated, not by the Equation (7), but by

$$b = \sqrt{\sum_i \frac{\{a_i\}^2}{c_i}}. \tag{19}$$

In this experiment, we use this modified method to calculate connection b to control the effective threshold.

4.4   Network with High Thresholds for S-cells

If the threshold of an S-cell is high, the tolerance area becomes small. In the vector space, reference vectors are created to cover all of the vectors of features contained in training patterns by those tolerance areas. Therefore, many reference vectors are required to cover because of the small tolerance area (see Figure 5(b)).

In the recognition phase, even if the input pattern indicated by x is slightly deformed from the learned one, which is denoted by X, the S-cell does not respond to x. Namely, the S-cell judges that x is in a different class from X, as the trained pattern.

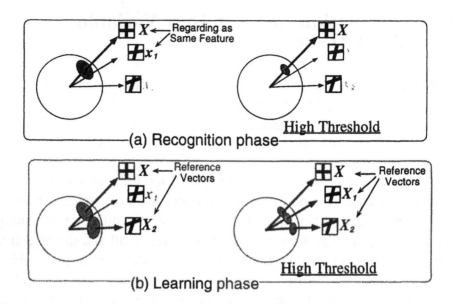

(a) Recognition phase

(b) Learning phase

Figure 5. When the threshold θ of the S-cell becomes high, the tolerance area becomes small. (a) In recognition phase, high threshold causes increase of selectivity. As a result, x_1 is not regarded as the same feature represented by X. (b) In the learning phase, high threshold causes a decrease in the size of the competition area. As a result, X_1 can become a reference vector of another S-cell because X and X_1 do not compete with each other.

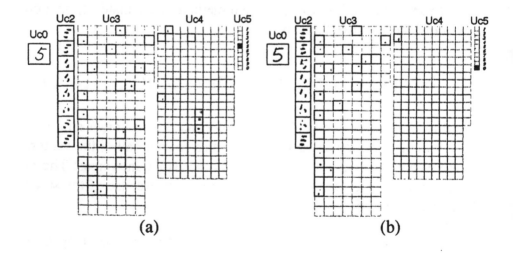

(a)   (b)

Figure 6. The response of a neocognitron with very high thresholds. Each rectangle showed C-cell-planes. Dark spots in the C-cell-planes show the response of cells. S-cell-planes are not shown in this figure. (a) The neocognitron correctly recognizes a trained pattern. (b) Even a slightly deformed pattern is classified erroneously, because the thresholds are too high. (modified from [2])

Figure 6 shows how the C-cells in the neocognitron respond when the thresholds of the S-cells are very high. Small rectangles stand for the cell-planes and dark spots in the cell-planes indicate the cells that fire. The darkness of the spot shows the firing strength. Figure 6(a) shows the response to one of the learned patterns. The pattern "5" is recognized correctly. Figure 6(b) shows the response to a slightly deformed input pattern. In U_{C4} layer, only a few cells respond. As a result, the recognition is failed. For this system, the pattern shown in (b) contains novel features. To cover these features by reference vectors, a lot of training patterns and learning are necessary. Of course, this neocognitron can correctly recognize some deformed patterns when those conditions are satisfied, but it requires a lot of cell-planes, connections, and learning time. Moreover, cell-planes are created for each deformed pattern in the training set, so that the generalization ability is very poor. Therefore, a too high threshold in the recognition phase is not practical, nor desirable.

4.5   Network with a Lower Threshold

If the threshold of an S-cell is low, the tolerance area becomes large (see Figure 7). In such a case, there is a risk that the important information would be lost in succeeding stages [2].

S-cells are reinforced by a kind of winner-take-all learning. The large tolerance area gives the chance that a single cell will always become a winner for many features, because the number of vectors falling in the tolerance area is proportional to the size of the tolerance area. In Figure 7(b), the low threshold causes the S-cell, whose reference vector is represented by X, to take the x_2 as well as the x_1 for learning. Hence, there is no chance for x_2 to become a reference vector of another S-cell. As a result of a winner-take-all learning, a lot of feature vectors are represented by a single reference vector. Therefore, only a small number of S-cells are created in the learning phase.

Figure 8 shows the neocognitron which is reinforced under a very low threshold condition. In U_{C3} of Figure 8, for instance, there are only four cell-planes. Thus, most of the important information has been lost in lower stages in the network. Figure 8(a) shows a response to a training

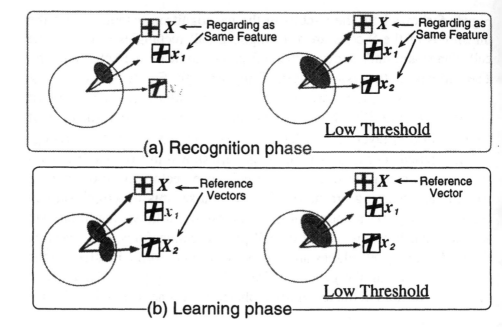

Figure 7. When thresholds become low, the tolerance area becomes large. (a) In the recognition phase, a more deformed pattern, such as x_2 regards the same feature represented by X. (b) In the learning phase, low threshold causes increase of the size of competition area. As a result, the reference vector X_2 is not created. The discrimination ability is deprived.

pattern "7", which is correctly recognized. In the case of Figure 8(b), however, the difference in shape between the pattern "7" and "9" are not detected in U_{C3}. The distribution of firing cells in the layer is very similar. As a result, the recognition failed to recognize the pattern.

A small threshold deprives selectivity in feature extraction and causes an S-cell to respond to different features which the cell has not yet learned. Generally speaking, this ability, called generalization ability, is necessary for pattern recognition. The problem of low threshold is that sufficient S-cell-planes are not created in the learning phase.

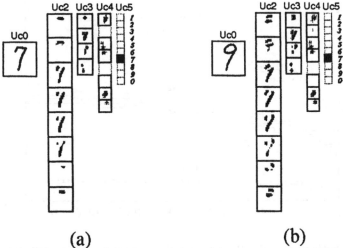

(a)   (b)

Figure 8. The response of a neocognitron with very low thresholds. Sufficient cell-planes are not created in the learning phase. (a) Response of the network to one of the training patterns. (b) Response to another pattern, which is erroneously recognized as "7". (modified from [2])

5 Different Thresholds in Learning and Recognition

The effect of different thresholds discussed in the previous section is summarized in Table 1. When a large threshold is used in the learning, a sufficient number of cell-planes are usually created. However, too large thresholds in the recognition phase cause the response of the neocognitron to over-fit to the learning pattern set, and the ability of generalization deteriorates.

Table 1. Summary of the effect of S-cell's threshold in learning phase and recognition phase.

	Low Threshold	High Threshold
Learning	Sufficient S-cells are NOT created.	Many (may be sufficient) feature-extracting cells are created.
Recognition	Good generalization ability.	Over-fitting to the learning pattern set.

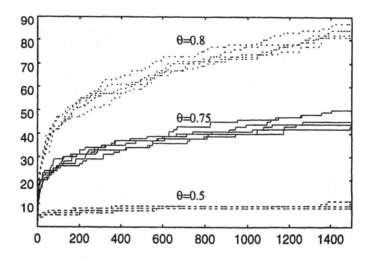

Figure 9. Number of S-cell-planes vs. the number of training patterns.

Low thresholds in recognition may produce a good generalization ability. Low thresholds in learning, however, do not produce enough feature extracting cell-planes. Therefore, we adopt a different threshold in the learning and recognition phases. In the learning phase, we use a high threshold so that many S-cell-planes are created. We use a lower threshold after finishing the learning, to ensure obtaining generalization ability.

Fukushima and Tanigawa have already shown that the recognition rate of the neocognitron can be increased when different threshold values are used in the learning and recognition, but only a small data set was used in their experiment [2].

5.1 Threshold in Learning Phase

In the learning phase, we need to set threshold θ at an appropriate value. If the training vector does not fall into the tolerance area of any S-cell, no S-cell responds and a new S-cell-plane is created because a competitive learning scheme is used. Therefore, the higher the threshold in the learning phase, the larger the number of S-cell-planes created. As a result, the network size becomes large.

In order to know the relationship between threshold θ and the number of S-cell-planes generation, we observed how the number of S-cell-planes in a layer U_{S3} changes with the threshold. Figure 9 shows the result. The abscissa is the number of training patterns and the ordinate is the number of S-cell-planes generated. We can see that the increase in the number of S-cell-plane is saturated. For example, when $\theta = 0.75$, the curve almost saturates at around 1,000 training patterns. For $\theta = 0.8$, however, the curve does not yet saturate at the point of 1,000 training patterns. If we use a lower θ, for example 0.5, the curve saturates before 1,000 patterns have been given.

When we have only 1,000 patterns available for training the network, what value of θ has to be used? If we have $\theta > 0.75$, the number of S-cell-planes does not reach a saturation level. We infer that the tolerance areas created by 1,000 training patterns are not enough to cover the feature vector space. Hence, when the number of training patterns is fixed to 1,000, we consider it better to set θ at 0.75 or lower. If $\theta < 0.75$, however, because θ is too low, 1,000 individual training patterns cannot be effectively used.

5.2   Upper Bound of the Threshold

In layers U_{C3}, U_{C4}, and U_{C5}, the cell density of each cell-plane is decreased from the preceding S-cell layer. From this structure constraint, the upper-bound of the thresholds appears.

We will illustrate this for the 1-dimensional neocognitron shown in Figure 10. To simplify the discussion, we assume here that the output of cells and connections take binary values $\{0, 1\}$, and C-cells in the U_{Cl} fires when the connected S-cell in U_{Sl} fires. Consider a situation where S-cell in layer U_{Sl} detects a feature and only that cell fires. In Figure 10(a), the S-cell 5 in U_{Sl} layer fires. The pattern appearing in layer U_{Sl} is spreading at layer U_{Cl}. As a result, three cells in U_{Cl} layer, such as C-cells 2, 3, and 4, fire. Here, assuming that the input connection of S-cells in U_{Sl+1} layer is $\{1, 1, 1\}$, then the S-cell 3 in U_{Sl+1} fires. The responses of other S-cells, such as 1, 2, 4, and 5, depend on the threshold.

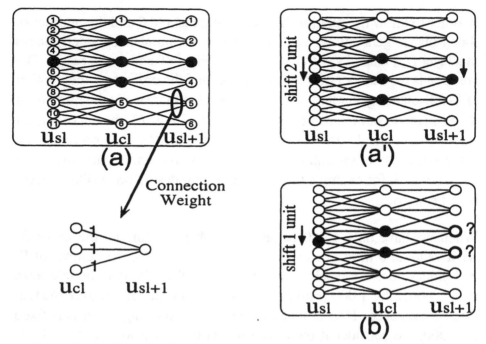

Figure 10. (a) When S-cell 5 (gray circle) in U_{Sl} layer fires, the spread pattern appears in U_{Cl} layer, that is, C-cells 2, 3, and 4 are fired. If the connection from U_{Cl} layer to U_{Sl+1} layer is $\{1, 1, 1\}$, the S-cell 3 in U_{Sl+1} responds. (a') When the pattern of U_{Sl} layer shifts downward by two units, and S-cell 7 becomes active, the pattern appearing in U_{Cl} layer shifts downward by one unit. As a result, S-cell 4 in U_{Sl+1} will fire. (b) If the pattern of U_{Sl} layer shifts downward by one unit, the pattern appearing in U_{Cl} layer is different from the pattern in (a) and (a'). The pattern appearing in U_{Cl} layer is not the same as the preferred feature for the S-cells 3 and 4 in U_{Sl+1}. Too large a threshold causes no response to a pattern shifted by one unit.

In Figure 10(a'), the pattern of U_{Sl} shifts downward by two units. Then, the pattern appearing in U_{Cl} shifts downward by one unit; i.e., C-cells 3, 4, and 5 in U_{Cl} fire. As a result, S-cell 4 in U_{Sl+1} fires.

Figure 10(b) shows the situation in which the pattern of U_{Sl} shifts downward by one unit. Only two cells in U_{Cl} layer (C-cells 3, and 4) will fire, and the pattern appearing in U_{Cl} is different from (a) and (a'). In this case, it is not the same as the preferred pattern for the S-cells 3 and 4 in U_{Sl+1} layer. Therefore, responses of S-cells in U_{Sl+1} layer depend upon its threshold. A large threshold of S-cells of U_{Sl+1} layer causes no response for a slightly shifted pattern. It is desirable that cells in U_{Sl+1} layer exhibit the same response to these shifted inputs.

Here we will use a mathematical analysis. We will use a vector notation to describe patterns of U_{Cl} in Figure 10(a) and Figure 10(b):

$$\begin{aligned}
\boldsymbol{x}_{(a)} &= (0, 1, 1, 1, 0, 0)^T \\
\boldsymbol{x}_{(b)} &= (0, 0, 1, 1, 0, 0)^T.
\end{aligned}$$

S-cell measures the distance of patterns by direction cosine (see Equation (1)). The direction cosine between $\boldsymbol{x}_{(a)}$ and $\boldsymbol{x}_{(b)}$ is

$$\frac{\boldsymbol{x}_{(a)} \cdot \boldsymbol{x}_{(b)}}{\|\boldsymbol{x}_{(a)}\| \cdot \|\boldsymbol{x}_{(b)}\|} = \frac{2}{\sqrt{6}} \sim 0.817. \tag{20}$$

In order to keep a shift invariance to these patterns, we have to set a threshold of θ_{l+1} of the S-cells, defined by Equation (17), in the range

$$\theta_{l+1} < 0.817. \tag{21}$$

When we set the threshold larger than this critical value, the neocognitron does not accept that shifted patterns are the same. This critical value depends on the shrinkage rate, that is, the ratio of cell densities between U_{Sl} layer and U_{Cl} layer, and the spreading of connections from U_{Sl} to U_{Cl}. When the shrinkage rate becomes large and the spreading of connections becomes small, this critical value becomes small.

Although we use 1-dimensional neocognitron in this analysis, the critical value of 2-dimensional neocognitron, such as used in this experiment, is lower than the analyzed value. In this experiment, the shrinkage rate between U_{S2} and U_{C2} is 3:1. To prevent this shift invariance in U_{C2}, we spread the size of the input connection area of C-cells. Moreover, this analysis assumes that a single S-cell fires in U_{Sl} layer; however, in U_{S2} layer (line extraction layer), neighboring S-cells have a tendency to fire because similar line components are gathered locally in the character pattern. Therefore, the estimated optimal value described in the next section is not so bad.

6 The Highest Stage; Classification Layer

S-cells in the U_{S5} layer are also trained by the unsupervised learning rule discussed in the previous section. The unsupervised learning in a training

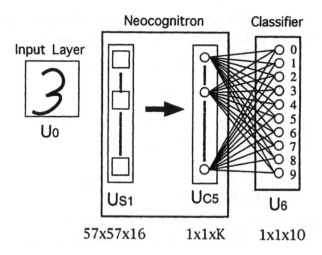

Figure 11. Simple perceptron-like layer for classification added after the conventical neocognitron.

pattern set consisting of a variety of deformed patterns usually crates a plural number of S-cell-planes for the patterns of the same category. Some patterns of the same category might elicit different responses from U_{S5}. These responses are transmitted to the U_{C5} layer whose cell-plane size is 1×1, that is, a single cell. For the purpose of pattern recognition, however, it is necessary to bind the responses of U_{C5} with digit symbols.

Therefore, we attached the simple-perceptron type network after the neocognitron system (Figure 11). The neocognitron system plays a role of pattern classification based on the shape of the patterns.

The perceptron-like layer binds symbol information with the feature information obtained by the neocognitron. The U_{C5} and the cells in perceptron-like layer are fully connected, and the connection weight values are randomly set in the initial state. The notation $w_{j\kappa}$ represents the connection between the κth cell in U_{C5} layer and the jth cell in the perceptron layer U_6. Therefore, index j is the digit category. The output of the jth cell in the classification layer U_6 is

$$y_j = f \left[\sum_{\kappa=1}^{K_{C5}+1} w_{j\kappa} u_{C5}(\kappa) \right].$$ (22)

To simplify the notation, we introduce the $(K_{C5} + 1)$th cell in U_{C5} layer whose output is always -1 ($u_{C5}(K_{C5} + 1) \equiv -1$). In other words, $w_{j\ K_{C5}+1}$ works as the threshold of the jth cell in U_6 layer. The learning of U_6 layer starts after finishing the learning of U_{S5} layer. The learning rule we adopt for this connection is Widrow-Hoff rule [10][11], which is a kind of supervised learning.

First, we define the cost function by

$$E[w] = \frac{1}{2} \sum_{j,\mu}^{P} (z_j^\mu - y_j)^2. \tag{23}$$

Here, P is the total number of training patterns labeled by μ. z_j^μ is the desired output, and y_j is the output of jth cell in classification layer U_6. We defined the output function as $f[x] = \tanh(x)$. To minimize $E[w]$ the updating value $\Delta w_{j\kappa}$ is derived

$$\Delta w_{j\kappa} = -\epsilon \frac{\partial E}{\partial w_{j\kappa}} = -\epsilon(z_j^\mu - y_j)(1 - y_j^2)u_{C5}(\kappa). \tag{24}$$

7 Results

Figure 12 shows an example of patterns in ETL-1 database. We prepare 3 sets of patterns randomly from this database. Those are 'training set', 'validation set', and 'test set'. These pattern sets are randomly sampled from the ETL-1 database, and are disjoint from each other.

The 'training set' consists of 1,000 patterns. We use this pattern set for creating connections in the neocognitron part and also for modifying the connections in the perceptron layer.

The 'validation set' also consists of 1,000 patterns. This pattern set is prepared for the optimization of the threshold in the neocognitron part. The optimization method will be described in Section 7.1. If we optimize thresholds using only the training set, the recognition rate for the training set may be improved. The generalization ability may be reduced. Therefore, we use another pattern set, that is 'validation set', for threshold optimizing as will be discussed later. This idea comes from "cross-validation"; that is, a statistical method [12] [13]. When we optimized

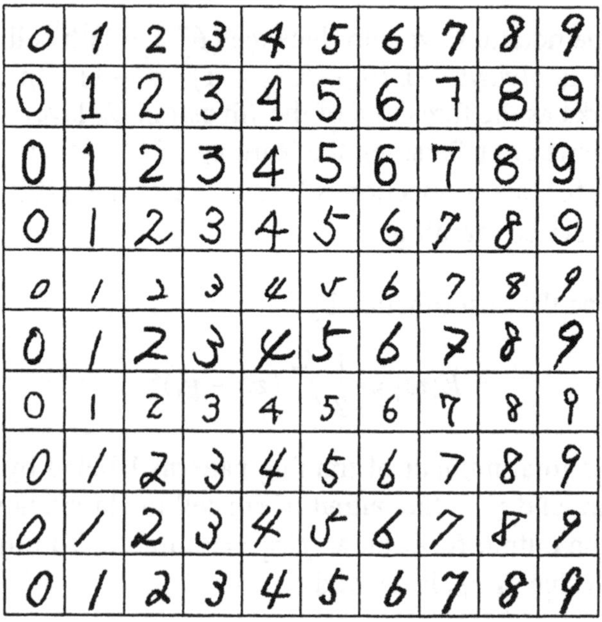

0	1	2	3	4	5	6	7	8	9
0	1	2	3	4	5	6	7	8	9
0	1	2	3	4	5	6	7	8	9
0	1	2	3	4	5	6	7	8	9
0	1	2	3	4	5	6	7	8	9
0	1	2	3	4	5	6	7	8	9
0	1	2	3	4	5	6	7	8	9
0	1	2	3	4	5	6	7	8	9
0	1	2	3	4	5	6	7	8	9
0	1	2	3	4	5	6	7	8	9

Figure 12. Example patterns in the ETL-1 database

threshold using only 'training set', training error, that is recognition error rate for training patterns, will be reduced; however, because the network may be over-fitted to the 'training set', the generalization error, that is recognition rate for novel patterns, will increase.

The last pattern set is a 'test set'. This is used for measuring the recognition rate of the network that has finished learning. Therefore, this pattern set must be novel to the system. We prepared 3,000 patterns for this set.

7.1 Optimization of Thresholds and Recognition Rate

For measuring the recognition rate, we must determine a threshold set. To obtain the best threshold set and to measure the recognition rate, we carried out the algorithm below.

Evaluation of recognition rate

Definition of notation:

θ_l^L: The threshold of the lth stage in the learning phase.

θ_R^L: The threshold of the lth stage in the recognition phase.

$\boldsymbol{\theta}$: A combination of threshold set, that is $(\theta_3^L, \theta_3^R, \theta_4^L, \theta_4^R, \theta_5^L, \theta_5^R)$.

Θ: Whole set of the $\boldsymbol{\theta}$.

1: Initialize variable for recognition rate R. $R = 0$
2: **repeat**
3:     Choose a $\boldsymbol{\theta}$ such that $\boldsymbol{\theta} \in \Theta$
4:     **for** $l = 3$ to 5 **do**
5:         Train the U_{Sl} layer with the thresholds θ_l^L by the training set. (Create and Modify connections of S-cells)
6:         Set the threshold of the U_{Sl} layer to θ_l^R.
7:     **end for**
8:     Train the perceptron-layer.
9:     Measure the recognition rate using the validation set, and record it as R.
10:    **if** $R > R_{\mathrm{opt}}$ **then**
11:        $\boldsymbol{\theta}_{\mathrm{opt}} = \boldsymbol{\theta}$
12:        $R_{\mathrm{opt}} = R$
13:    **end if**
14: **until** whole threshold sets in Θ are examined.
15: Measure the recognition rate of the neocognitron with threshold $\boldsymbol{\theta}_{\mathrm{opt}}$ by test set.

Reinforcement is performed from lower to higher stages. When the lth S-cell layer has finished training, the threshold parameter is set to θ_l^R.

We measured the recognition rate for many different combinations of the threshold values. 1,000 training patterns were given 5 times for the training of each layer during the learning phase. We searched the best combi-

Table 2. The best threshold set θ_{opt}

	$l = 3$	$l = 4$	$l = 5$
θ_l^L	0.75	0.65	0.76
θ_l^R	0.66	0.53	0.45

nation of thresholds, which produces the maximum recognition rate for the validation pattern set.

When we used the threshold parameters, that is, θ_{opt} in Table 2, the recognition rate for the validation set was 98.4 % and 100.0 % for the training set. Finally, we obtained 98.1 % for the test set.

8   Summary

In this chapter, we discussed pattern recognition by the neocognitron with dual threshold. Using a large scale database, such as ETL-1, we have demonstrated the effectiveness of adopting dual threshold values.

The only problem of the threshold optimization method is the long computational time. If we change the thresholds of the lth stage, the responses of the layer vary. Thus, we have to create and modify the connection weight again after $(l + 1)$ stages. A change of threshold θ_l^R or θ_l^L causes a drastic change of structures, that is, the number of cell-planes, connections, and so on, of the succeeding layers. However, once optimal thresholds have been given, the neocognitron can quickly learn and recognize.

Acknowledgments

We thank Ken-ichi Nagahara for helpful discussions and computer simulation programming. This research was supported in part by Grants-in-Aid #09308010, #10164231, #08780358, and #10780231 for Scientific Research from the Ministry of Education, Science,

Sports and Culture of Japan; and by a grant for Frontier Research Projects in Telecommunications from the Ministry of Posts and Telecommunications of Japan. We thank ETL (Electrotechnical Laboratory, Agency of Industrial Science and Technology, Ministry of Internatinal Trade and Industry) for giving us permission to use the ETL-1 database.

References

[1] Fukushima, K. (1980), "Neocognitron: A self-organizing neural network model for a mechanism of pattern recognition unaffected by shift in position," *Biological Cybernetics*, **36**(4):193–202.

[2] Fukushima, K. and Tanigawa, M. (1996), "Use of different thresholds in learning and recognition," *Neurocomputing*, **11**(1):1–17.

[3] LeCun, Y. et al. (1989), "Backpropagation applied to handwritten zip code recognition," *Neural Computation*, **1**(4):541–551.

[4] Hubel, D.H. and Wiesel, T.N. (1977), "Functional architecture of macaque monkey visual cortex," *Proceedings of Royal Society of London*, **198**(1130):1–59, July.

[5] Hubel, D.H. and Wiesel, T.N. (1959), "Receptive fields of single neurones in the cat's striate cortex," *J.Physiol.(Lond.)*, 148:574–591.

[6] Fukushima, K. and Wake, N. (1992), "An improved learning algorithm for the neocognitron," in I. Aleksander J. Taylor (Ed.), *Artificial Neural Networks*, Vol. 1 of 2, pp. 497–505. Amsterdam: North-Holland.

[7] Fukushima, K. (1988), "Neocognitron: A hierahical neural network capable of visual pattern recognition," *Neural Networks*, 1:119–130.

[8] Fukushima, K. (1989), "Analysis of the process of visual pattern recognition by the neocognitron," *Neural Networks*, 2:413–420.

[9] Ohno, M., Okada, M. and Fukushima, K. (1995), "Neocognitron learned by backpropagation," *Systems and Computers in Japan*, **26**(5):19–28.

[10] Widrow, B. and Hoff, M.E. (1960), "Adaptive switching circuits," *1960 IRE WESCON Convention Record*, Part 4, pp. 96–104.

[11] Rumelhart, D.E., McClelland, J.L. and PDP Research Group (1986), *Parallel Distributed Processing: Explorations in Microstructure of Cognition*, Volume 1. MIT Press.

[12] Bishop, C.M. (1995), *Neural Networks for Pattern Recognition*. Oxford University Press.

[13] Stone, M. (1978), "Cross-validation: A review," *Math. Operations Stat. Ser. Stat*, **9**(1):127–139.

Chapter 3:

Recognition of Rotated Patterns Using a Neocognitron

RECOGNITION OF ROTATED PATTERNS USING A NEOCOGNITRON

S. Satoh
Aso Laboratory, Department of Electrical Communications
Tohoku University, Aoba-yama 05, Sendai 980-8579, Japan
e-mail: shun@aso.ecei.tohoku.ac.jp

J. Kuroiwa
The Division of Mathematical and Information Sciences,
Hiroshima University
Higashi-Hiroshima 739-8521, Japan

H. Aso
Aso Laboratory, Department of Electrical Communications
Tohoku University, Aoba-yama 05, Sendai 980-8579, Japan

S. Miyake
Department of Applied Physics, Tohoku University
Aoba-yama 04, Sendai 980-8579, Japan

A neocognitron is a neural network model which is considerably robust for distortion, scaling, and/or translation of patterns. However, it cannot recognize largely rotated patterns. A rotation-invariant neocognitron is constructed by extending the neocognitron which can recognize translated, scaled and/or distorted patterns from training ones. The rotation-invariant neocognitron is proposed in order to support the weak point of the original neocognitron. Numerical simulations show that the rotation-invariant neocognitron can correctly recognize input patterns without being affected by global and/or local rotation as well as translation, scaling and distortion from training patterns.

0-8493-9807-X/99/$0.00+$.50

1 Introduction

Many recognition systems which are not affected by distortion in shapes of patterns or shifts in position have been proposed that are expected to be robust for distortions of a standard pattern [1] [2] [3]. However, they are not necessarily satisfactory in the sense that they cannot recognize rotated, distorted and shifted patterns. For example, some models can recognize distorted patterns but cannot recognize rotated ones. It is very important to realize reliable recognition systems which are insensitive to rotation as well as scaling, translation and distortion.

A neocognitron [4] is a multilayered neural network model for pattern recognition. The network has a structure based on the hierarchical model by Hubel and Wiesel [5], and cells in a higher layer of the neocognitron are more insensitive to shifts in positions of input patterns and have a tendency to respond to more complex features of patterns. Although the hierarchical model by Hubel and Wiesel is not necessarily supported by physiological experiments, we can employ the hierarchical structure as a possible model which realizes the function of the integration of primary visual information [6]. The neocognitron is an interesting model which has the function of the integration of visual information. The model can acquire an extremely robust ability of pattern recognition through an unsupervised learning. The robustness is for distortions in shapes of patterns and shifts in those positions and noise. The model is tested by using handwritten character database, ETL-1 database,[1] and shows the recognition rate of more than 98% by attaching additional layers to the neocognitron proposed in [7] [8]. However, it cannot recognize largely rotated patterns.

One can correctly recognize rotated patterns in an instant if the patterns are comparatively simple. This shows that rotated patterns can be recognized by a bottom-up type model such as a neocognitron. We propose a new hierarchical network model for pattern recognition, called *a rotation-invariant neocognitron*, which is also insensitive to rotations of patterns as well as distortions, scalings and/or translations of patterns.

---

[1] A character database published by Electro-technical Laboratory, Japan.

2 Rotation-Invariant Neocognitron

2.1 Short Review of Neocognitron

We briefly explain the structure and function of the original neocognitron proposed by Fukushima [4]. The neocognitron is composed of cascaded connections of a number of modular structures, U_{Sl} and U_{Cl}, preceded by an input layer U_0 consisting of photoreceptor array as shown in Figure 1. Here U_{Sl} denotes a layer consisting of S-cells in the lth module,

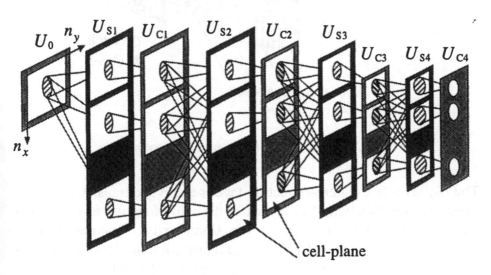

Figure 1. A basic structure of neocognitron.

and U_{Cl} a layer consisting of C-cells in the lth module. Each layer is composed of a number of cell-planes. A cell-plane contains many S-cells or C-cells, and is represented by a white rectangle in Figure 1. The number of cells in each cell-plane is designed to decrease as a number of module, l, increases. An S-cell in U_{Sl}, except in U_{S1}, has two kinds of connections; excitatory plastic connections with C-cells in U_{Cl-1} layer [2] and an inhibitory plastic connection with a V-cell in U_{Sl} layer. V-cells and inhibitory connections are not depicted in Figure 1 for simplicity. A C-cell has fixed connections with S-cells in U_{Sl} layer (see Figure 1). Plastic connections are reinforced using unsupervised learning while training patterns are presented at U_0.

---

[2]U_{C0} layer stands for U_0 layer.

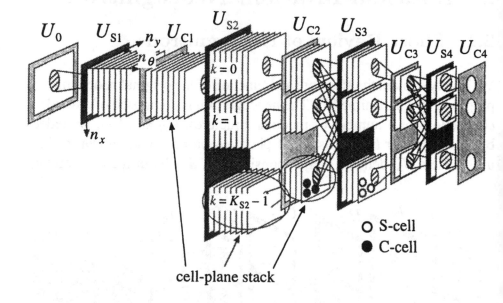

Figure 2. A structure of rotation-invariant neocognitron.

The U_{Sl} layer is a feature-extraction layer sensitive to translation of features, and the following U_{Cl} layer has the same function as U_{Sl}, but is insensitive to translation and distortion. Complexity of features to be detected and insensitivity to translation and distortion increase as the order of the module increases by cascaded connection of U_S and U_C layers. U_{C4} layer at the highest module is a recognition layer, which represents the result of pattern recognition. The neocognitron is insensitive to scaling, translation and distortion, and these functions are realized by a blurring operation of C-cells. The latest neocognitron [7] [8] has an additional module, an edge-detecting layer or a bend-detecting layer to improve the recognition rate, but those are not depicted in Figure 1.

2.2   Structure of Rotation-Invariant Neocognitron

A rotation-invariant neocognitron [9] [10] is also composed of cascaded connections of a number of modular structures, U_{Sl} and U_{Cl}, preceded by an input layer U_0, as shown in Figure 2. A layer is composed of a

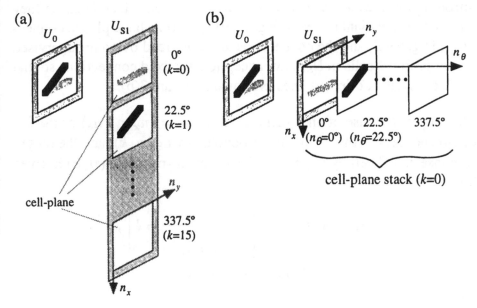

Figure 3. (a) Detection of an oriented line by cell-planes in original neocognitron. (b) Detection of an oriented line by a cell-plane stack in rotation-invariant neocognitron.

number of *cell-plane stacks*, and different cell-plane stacks detect different features of inputs. A cell-plane stack is composed of a number of cell-planes, and each cell-plane detects a different rotation angle of the features. Each cell in a cell-plane stack is located in a *three*-dimensional space. In our model the information of the orientation of an input pattern is represented by a number n_θ assigned to a cell-plane in a cell-plane stack, and the information of the position of the pattern is represented by coordinates (n_x, n_y) of a firing cell in the cell-plane of n_θ. While in the original neocognitron each rotation-invariant pattern with a different orientation is detected in a different cell-plane, in the rotation-invariant neocognitron such rotation-invariant pattern is detected by an S-cell in a same cell-plane stack. The difference between a cell-plane and a cell-plane stack in U_{S1} is shown in Figure 3. Figure 3(a) shows the response of U_{S1} in the original neocognitron when $\angle$-shaped pattern is presented on U_0, and Figure 3(b) shows the response of U_{S1} in the rotation-invariant neocognitron. In the first module, U_{S1} and U_{C1}, both the original neocognitron and the rotation-invariant neocognitron detect oriented lines. In the original neocognitron each line with a different orientation is detected in a different cell-plane (Figure 3(a)). In the rotation-invariant neocognitron, those lines are detected in a same cell-plane stack because they are rotation-invariant patterns (Figure 3(b)). S-cells have plastic connections, in general, and the connections are reinforced by an unsupervised learning. But S-cells in U_{S1} have exceptionally fixed connections so that the S-cells detect oriented lines as shown in Figure 3(b).

The output response of an S-cell located on $n = (n_x, n_y, n_\theta)$ of the kth cell-plane stack in the lth module is denoted by $u_{Sl}(n, k)$, and the output response of a C-cell by $u_{Cl}(n, k)$. The output response $u_{Sl}(n, k)$ is given by

$$u_{Sl}(n, k) = r_l \cdot \phi \left[\frac{1 + e}{1 + \dfrac{r_l}{1 + r_l} \cdot i} - 1 \right], \tag{1}$$

$$\phi(x) = \max(x, 0), \tag{2}$$

where

$$e = \sum_{\kappa=0}^{K_{Cl-1}-1} \sum_{\nu \in A_l} a_l(\nu, n, \kappa, k) \cdot u_{Cl-1}(n \underset{T_{Cl-1}}{\oplus} \nu, \kappa), \tag{3}$$

$$i = b_l(k) \cdot u_{Vl}(n). \tag{4}$$

A binomial operator $\underset{M}{\oplus}$ with M is defined by

$$\begin{cases} n_x \underset{M}{\oplus} \nu_x & \overset{\text{def}}{=} & n_x + \nu_x, \\ n_y \underset{M}{\oplus} \nu_y & \overset{\text{def}}{=} & n_y + \nu_y, \\ n_\theta \underset{M}{\oplus} \nu_\theta & \overset{\text{def}}{=} & (n_\theta + \nu_\theta) \bmod M. \end{cases} \tag{5}$$

Here r_l denotes the threshold value of an S-cell in the lth layer, $a_l(\nu, n, \kappa, k)$ represents an excitatory connection from a C-cell in the previous module to the S-cell and $b_l(k)$ an inhibitory connection from a V-cell to an S-cell —V-cells send inhibitory inputs to S-cells and are not depicted in Figure 2 for simplicity. Each connection is linked to a restricted number of C-cells in the preceding module, and A_l defines the restricted number of C-cells. K_{Cl} denotes the number of cell-plane stacks in the U_{Cl} layer, and T_{Cl} the number of cell-planes in the U_{Cl} layer. For instance, $T_{C1} = 8$ means that the orientation of a pattern detected in U_{C1} is quantized into eight points, $0°, 45°, \cdots, 315°$.

The output response of a V-cell is given by

$$u_{Vl}(n) = \sqrt{\sum_{\kappa=0}^{K_{Cl-1}-1} \sum_{\nu \in A_l} c_l(\nu) \cdot \left\{ u_{Cl-1}(n \underset{T_{Cl-1}}{\oplus} \nu, \kappa) \right\}^2} \qquad (6)$$

where $c_l(\nu)$ is an excitatory connection from a C-cell to a V-cell, which takes a fixed value during learning.

The output response of a C-cell is given by

$$u_{Cl}(n, k) = \psi \left[\sum_{\nu \in D_l} d_l(\nu) \cdot u_{Sl}(n \underset{T_{Sl}}{\oplus} \nu, k) \right], \qquad (7)$$

where the function ψ is defined by

$$\psi(x) = \frac{\phi(x)}{\phi(x) + 1}. \qquad (8)$$

Here $d_l(\nu)$ is an excitatory connection from an S-cell to a C-cell, $D_l (\subset Z^3)$ represents a restricted region of the connection, and T_{Sl} a number of cell-planes in the U_{Sl} layer.

At learning stages, excitatory connections $a_l(\nu, n, \kappa, k)$ and inhibitory connections $b_l(k)$ are modified.

2.3 Unsupervised Learning and Detection of Rotated Patterns

An unsupervised learning has been performed step by step from lower modules to higher modules. That is, the training of a higher module is performed after completion of the training of the preceding module. We divide an unsupervised learning process into two stages, STAGE(I) and STAGE(II). STAGE(I): a winner-cell (called a *seed-cell*) of S-cells denoted by $U_{Sl}(\hat{n}, \hat{k})$ is chosen, and then excitatory connections of the winner-cell $a_l(\nu, \hat{n}, \kappa, \hat{k})$ and inhibitory connections $b_l(\hat{k})$ are modified by unsupervised learning using training patterns. STAGE(II): the other excitatory connections from C-cells to S-cells at all the $n \neq \hat{n}$ and on

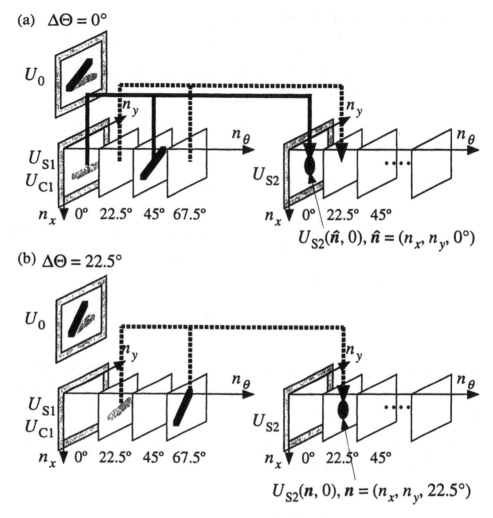

Figure 4. A response of U_{S2} layer. $\Delta\Theta$ is a shift of angle compared with a learning pattern. C-cells will also fire in practice, but the firing is omitted for simplicity.

the $k = \hat{k}$ cell-plane stack are modified by the same amount and by the same spatial distribution as connections of the seed-cell.

Now let us consider the functions of U_{S2} layer as an example to explain how S-cells detect rotated patterns after completion of unsupervised learning, which are schematically shown in Figure 4. We consider the situation where a ∠-shaped pattern is presented on U_0. At STAGE(I), if $U_{S2}(\hat{n}, 0), \hat{n} = (n_x, n_y, 0°)$ is chosen as a seed-cell for the pattern, the excitatory connection of the seed-cell, $a_2(\boldsymbol{\nu}, \hat{\boldsymbol{n}}, 0, 0)$, is modified as a thick arrow shown in Figure 4(a). The inhibitory connection of the seed-cell, $b_2(0)$, is also modified but it is not depicted for simplicity. At STAGE(II), the other connections of S-cells on the $\hat{k} = 0$ cell-planes are reinforced. A striped arrow shows a distribution of reinforced $a_2(\boldsymbol{\nu}, \boldsymbol{n}, 0, 0)$ with $\boldsymbol{n} = (n_x, n_y, 22.5°)$ corresponding to the cell-plane of $n_\theta = 22.5°$. Other connections of $n_\theta = 45°, 67.5°, \cdots, 337.5°$ are also modified in the same manner. If a rotated $\angle$-shaped pattern of $22.5°$, which is not used in the training phase, is presented on U_0 layer, the S-cell corresponding to the angle, $U_{S2}(\boldsymbol{n}, 0)$ with $\boldsymbol{n} = (n_x, n_y, 22.5°)$, fires as shown in Figure 4(b). Thus the rotated pattern of n_θ is detected by an S-cell in the cell-plane of n_θ.

The plastic connections of a seed-cell are modified by

$$\Delta a_l(\boldsymbol{\nu}, \hat{\boldsymbol{n}}, \kappa, \hat{k}) = q_l \cdot c_l(\boldsymbol{\nu}) \cdot u_{\mathrm{C}l-1}(\hat{\boldsymbol{n}} + \boldsymbol{\nu}, \kappa), \tag{9}$$
$$\Delta b_l(\hat{k}) = q_l \cdot u_{\mathrm{V}l}(\hat{\boldsymbol{n}}), \tag{10}$$

where q_l is a constant which controls a learning efficiency, and $\hat{\boldsymbol{n}}$ denotes a location of the seed-cell.

2.4 Robustness for Rotations

C-cells of the original neocognitron have a two-dimensional blurring operation for the firing pattern of S-cells in the same module [11]. The operation makes a C-cell less sensitive to feature detection than the S-cell. We also introduce the blurring operation in the rotation-invariant neocognitron. In our model the blurring operation by C-cells is extended to the *three*-dimensional space, and C-cells become less sensitive to rotations as well as positions of patterns. Such operations are executed in every $U_{\mathrm{C}l}$ layer, and C-cells in the last layer acquire the ability to detect globally and locally rotated patterns from training patterns.

A mechanism of robustness for the distortions and/or rotations of input patterns is schematically shown in Figure 5, where a certain S-cell in U_{S3} is trained by a standard pattern "A", which is composed of three features, presented on the input layer U_0. In Figure 5(a), a shaded circle represents a range of tolerance for translations of inputs and a double-sided arrow a range of tolerance for rotation of inputs. These ranges are defined by the three-dimensional blurring operations by C-cells in the U_{C2} layer, and equal to D_2. This S-cell detects any distorted and/or rotated pattern from

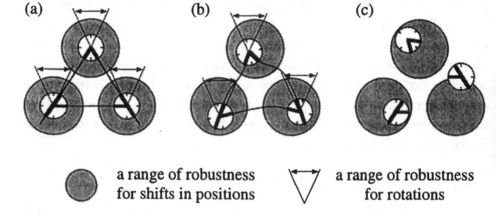

(a)   (b)   (c)

a range of robustness for shifts in positions   a range of robustness for rotations

Figure 5. Diagram illustrating a mechanism of tolerance for distortions and rotations.

the standard pattern "A", as long as the deviations of each feature fall into two kinds of ranges as shown in Figure 5(b). However, the S-cell doesn't fire for an extremely deformed or rotated pattern as shown in Figure 5(c) because the deviations exceed the ranges. In the rotation-invariant neocognitron, each S-cell can detect locally rotated patterns because of the extended blurring operation.

$$2 \quad 3 \quad 4 \quad 5 \quad 6 \quad 7$$

Figure 6. Training patterns represented on U_0 layer.

3   Simulation

We examine the recognition ability of the rotation-invariant neocognitron for scaled, translated, distorted and/or rotated patterns. The model is trained by some numerical characters as shown in Figure 6, and rotated and/or distorted patterns are used for tests after completion of a learning.

The parameters about the number of cells are given in Table 1. A size of cell-plane stack in U_{Sl} is $N_{Sl} \times N_{Sl} \times T_{Sl}$, and $N_{Cl} \times N_{Cl} \times T_{Cl}$ in U_{Cl}. Numbers marked by asterisks in columns of Table 1 are not defined, and numbers in parentheses in columns are after completion of learning because these vary during the learning. The sizes of the areas, A_l and D_l, are given in Table 2. The values of r_l and q_l are given in Table 3.

Table 1. Number of cell-plane stack and number of cells in one cell-plane stack. Numbers in columns marked by asterisks are not defined and numbers in parentheses in columns are after completion of learning.

	$l = 0$	$l = 1$	$l = 2$	$l = 3$	$l = 4$
K_{Sl}	*	1	(5)	(7)	(6)
N_{Sl}	*	59	17	13	3
T_{Sl}	*	16	8	4	2
K_{Cl}	1	1	(5)	(7)	(6)
N_{Cl}	61	19	17	10	1
T_{Cl}	1	8	4	2	1

Table 2. Size of area connected with one cell. A_l and D_l denote the size connected with an S-cell and a C-cell, respectively.

	$l = 1$	$l = 2$	$l = 3$	$l = 4$
A_l	$3 \times 3 \times 1$	$3 \times 3 \times 8$	$5 \times 5 \times 4$	$3 \times 3 \times 2$
D_l	$5 \times 5 \times 3$	$3 \times 3 \times 3$	$5 \times 5 \times 3$	$3 \times 3 \times 2$

Table 3. Values of r_l and q_l.

	$l = 2$	$l = 3$	$l = 4$
r_l	2.3	2.2	2.6
q_l	1.0×10^5	1.0×10^5	1.0×10^5

The robustness of the original neocognitron and the rotation-invariant neocognitron for rotations are shown in Figure 7. It is shown that the rotation-invariant neocognitron can recognize arbitrarily rotated patterns. This complete insensitiveness to rotations is very useful in many fields, e.g., in document-analysis or in automatic identification of parts of a machine, where recognition of arbitrarily rotated patterns is required.

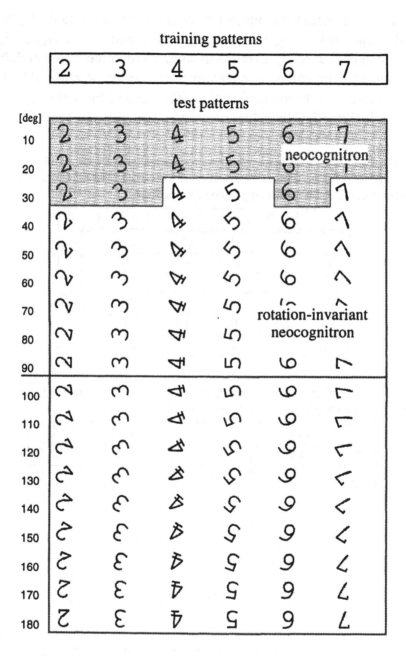

Figure 7. Correctly recognized patterns by the original neocognitron and the rotation-invariant neocognitron.

We can see that the rotation-invariant neocognitron is robust for rotations and/or distortions as shown in Figure 8(a). It is surprising that the rotation-invariant neocognitron can also recognize patterns in which some parts of each training pattern are rotated in different angles as shown in Figure 8(b). Patterns in Figure 8(b) are not globally rotated, but cannot be recognized by the neocognitron. This result indicates that our model is better in robustness for distortions than the original neocognitron. We conclude that our model is very useful for recognition

(a)

(b)

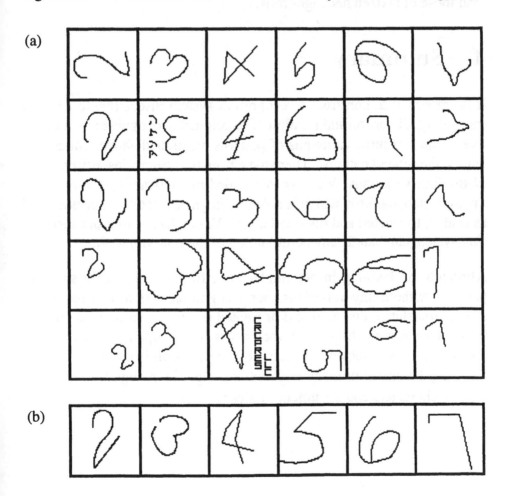

Figure 8. Examples of correctly recognized patterns. (a) Rotated and/or distorted patterns. (b) A sample of patterns which rotate some parts of each training pattern.

of handwritten characters which contain many features that are locally rotated.

If one uses sixteen original neocognitrons corresponding to different rotation-angles in order to recognize rotated patterns, one would need to prepare many rotated patterns and execute a learning for each neocognitron. On the other hand, our model needs only standard patterns for learning, and the total number of cells and connections are much less than those of sixteen neocognitrons.

4 Summary

It is important to construct a recognition system insensitive to scaling, translating, distortion and rotation. The original neocognitron is insensitive to the first three variations of patterns mentioned above, but cannot handle the rotated patterns. There have been no systems insensitive to all of the four variations. We have succeeded in constructing the rotation-invariant neocognitron which is able to recognize rotated patterns as well as scaled, translated and distorted ones. We emphasize that our recognition system can handle any combination of the four variations.

Although the learning in our model is performed using only standard patterns without any variations, our recognition system can recognize scaled, translated, distorted and/or rotated patterns in numerical simulations. Our system shows high robustness for recognition of numerical characters. This robustness means that the rotation-invariant neocognitron may be useful in many applications, e.g., the recognition of handwritten characters and document analysis.

References

[1] M. Fukumi, S. Omatu, and Y. Nishikawa, (1997), "Rotation-Invariant Neural Pattern Recognition System Estimating a Rotation Angle," *IEEE Trans., Neural Network*, Vol. 8, pp. 568–581.

[2] M. B. Reid, L. Spirkovska, and E. Ochoa, (1989) "Rapid training of higher order neural networks for invariant pattern recognition," *Proc. Int. Joint Conf. Neural Networks*, Vol. 1, pp. 689–692.

[3] B. Widrow, R. G. Winter, and R. A. Baxter, (1988), "Layered neural nets for pattern recognition," *IEEE Trans. Acoust., Speech, Signal Processing*, Vol. ASSP-36, pp. 1109–1118.

[4] K. Fukushima, (1988), "Neocognitron: A hierarchical neural network capable of visual pattern recognition," *Neural Networks*, Vol. 1, No. 2, pp. 119–130.

[5] D. H. Hubel and T. N. Wiesel, (1965), "Receptive fields and functional architecture in two nonstriate visual area (18 and 19) of the cat," *J. Neurophysiol.*, Vol. 28, pp. 229–289.

[6] K. Fukushima, (1980), "Neocognitron: A self-organizing neural network model for a mechanism of pattern recognition unaffected by shift in position," *Biol. Cybernetics*, Vol. 36, pp. 193–202.

[7] K. Fukushima and N. Wake, (1992), "Improved Neocognitron with Bend-Detecting Cells," International Joint Conference on Neural Networks IJCNN'92, Vol. 4, pp. 190-195.

[8] H. Shouno, K. Nagahara, K. Fukushima, and M. Okada, (1997), "Neocognitron applied to hand-written Digit Recognition — Evaluation with large Character Database—(in Japanese)," *IEICE Tech. Rep.*, NC97-19, pp. 65–71.

[9] S. Satoh, J. Kuroiwa, H. Aso, and S. Miyake, (1998), "A rotation-invariant Neocognitron (in Japanese)," *IEICE Trans.*, J81–D–II, No. 6, pp. 1365–1373.

[10] S. Satoh, J. Kuroiwa, H. Aso, and S. Miyake, (1997), "Recognition of rotated patterns using neocognitron," Proceedings of the International Conference on Neural Information Processing (ICONIP'97), Vol. 1, pp. 112–116.

[11] K. Fukushima, (1989), "Analysis of the process of visual pattern recognition by the neocognitron," *Neural Networks*, Vol. 2, pp. 413–420.

[12] K. Fukushima and N. Wake, (1992), "An improved learning algorithm for the neocognitron," *ICANN'92*, Brighton U.K., pp. 4–7, Sept.

Chapter 4:

A Soft Computing Approach to Handwritten Numeral Recognition

A SOFT COMPUTING APPROACH TO HANDWRITTEN NUMERAL RECOGNITION

J.F. Baldwin, T.P. Martin and **O. Stylianidis**
Artificial Intelligence Group, Dept. of Engineering Mathematics
University of Bristol, Bristol BS8 1TR, U.K.

Soft computing is a key technology for the management of uncertainty, encompassing a range of techniques including fuzzy methods, neural networks and probabilistic reasoning. In the work reported here, we integrate these different aspects of soft computing to develop a system for recognizing handwritten numerals incorporating off-line learning and feature discovery. The system makes use of Kohonen's *Self-Organizing Maps* to invent features that may describe the digits, and Baldwin's *Evidential Support Logic Programming* (in Fril) to create rules and reason with the selected features.

Characters are normalized to an 8×8 bitmap and subsequently split into 4 regions. The region bitmap patterns constitute the input to a Kohonen self-organizing network, generated by a growth process. The result is a dimensionality-reducing and topology-preserving mapping, where each output node is similar to what we would perceive as lines and curves in each region. We use fuzzy sets, extracted from this data to represent sets of feature values and provide a mechanism for generalization. Features are combined by means of the evidential logic rule. We employ semantic unification, which is based on Baldwin's mass assignment theory, to determine the correspondence between the observed character and the generalized prototypical case as described by the evidential logic rule. This produces a support for the digit belonging to each class, used in assessing the classification of each input pattern. The system is implemented in Fril.

In our investigation, we use the CEDAR database of handwritten numerals collected from postal addresses. We compare the system's performance to existing architectures used for the same purpose.

0-8493-9807-X/99/$0.00+$.50

1   Introduction

Soft computing encompasses a range of techniques, including fuzzy logic, neural network theory, and probabilistic reasoning – the distinguishing feature being a tolerance of imprecision, uncertainty and partial truth [1]. In this chapter, we use a combination of the main aspects of soft computing to create classifiers. A Kohonen-style self-organizing net is used to generate new features; fuzzy sets are used to summarize feature values from a training database of examples; and evidential support logic rules are used to classify test data.

The conventional statistical approach to performing classification is to use a discriminant classifier that constructs boundaries which discriminate between objects of different categories. This is not an easy problem – especially if we consider that there may be considerable overlap among the categories. An alternative approach is to consider degrees of membership that an object has in each available category and then decide to which class it should be assigned.

Our chosen classification problem is that of unconstrained handwritten digit recognition. Compared to general alphanumeric character recognition, this is an easier task as it deals with only ten different classes rather than all 62 alphanumerics. However, it retains most of the problems associated with variability in shape and form, overlapping classes and noise.

Handwritten character recognition systems have been proposed and implemented in a number of different ways. In order to maximize the performance of unconstrained handwritten numeral recognition, the following two aspects should be considered; one is the design of a feature extractor that produces discriminating features, and the other is the design of a classifier which has good generalization power.

Our work has been motivated by these two issues. The proposed scheme consists of two stages: an automatic feature extraction preprocessor based on a neural net and a classifier based on evidential support logic [2], [3].

Neural networks have been intensively studied in recent years in an effort to achieve human-like performance in recognition and classification tasks. The self-organizing map (SOM) proposed by

Kohonen [4] is one of the network structures widely used in pattern recognition. The SOM is a sheet-like network in which neighboring units compete in their activities by means of their mutual lateral interaction and become specifically tuned to various input patterns. Following training, weight patterns from the inputs to each node can be considered as prototypes which are used to classify patterns by means of a nearest neighbor classifier.

Using features discovered by the SOM, evidential rules are then generated by learning from training patterns. Evidential support logic is a method of building up support for a hypothesis from the weighted support of a series of features associated with the hypothesis. The known examples must be generalized somehow to account for unseen data. This is achieved through the use of fuzzy sets extracted from the features.

Finally, we demonstrate how utilization of structural relationships among these features can improve the recognition accuracy within the framework of evidential logic programming.

2 Description of Data and Preprocessing

The starting point for our investigation was a database of handwritten alphanumeric characters compiled by the Center of Excellence for Document Analysis and Recognition (CEDAR) at the State University of New York at Buffalo. Because the characters were scanned from "live" mail, the data were unconstrained for writer, style and method of preparation. These characteristics help overcome the limitations of databases that contain only isolated characters or have been prepared in a laboratory setting under prescribed circumstances. Furthermore, by using the CEDAR database rather than a small laboratory database, our work will yield results directly comparable to the work of other researchers in the field. Only a few important aspects of the database will be presented here; for full details see [5].

The characters in the CEDAR database were extracted from entire address blocks. The assignment of the "correct" truth values to these data was performed by a procedure that extracted connected components from an address image and displayed them in isolation to an operator. A truth value was assigned to a component if its truth was obvious in isolation. If the truth was not obvious, the operator used the coordinates of the component to locate it in the original address block. The surrounding context of the component was then used to assign the truth value. The database is explicitly split into training and test sets by randomly choosing approximately 10% of the available images as the test set.

We have chosen to work with the subset of numerical characters. The breakdown into training and testing sets is shown in Table 1.

Table 1. Numeric Character Set Distribution.

DataSet	Class										
	0	**1**	**2**	**3**	**4**	**5**	**6**	**7**	**8**	**9**	**Total**
Training	811	1160	778	467	607	342	469	429	346	393	*5802*
Testing	102	136	103	68	63	·41	47	48	46	53	*707*

In order to eliminate variations in the thickness of the lines, we applied a thinning algorithm [6] to the characters. We subsequently normalized the resulting bitmap to an 8×8 grid, to minimize the effect of image size on the recognition process. Figure 1 shows examples of digit images as they were found in the database and Figure 2 an example of a digit before and after normalization.

3   Extracting Rules from Data

The methods used in forming predictive rules from the data rely on the theoretical framework provided by mass assignment theory. This enables us to relate probabilistic and fuzzy uncertainty, and (for example) to generate fuzzy sets from statistical data. Brief details of mass assignment theory are given below, to make this chapter fairly self-contained.

3.1 Essentials of Mass Assignment Theory

Let us consider that a dice is thrown and the value is *small* where *small* is a fuzzy set defined as

$$small = 1 / 1 + 2 / 0.9 + 3 / 0.4$$

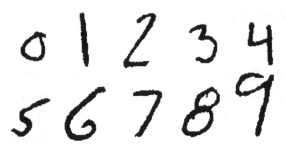

Figure 1. A random selection of handwritten digits from the CEDAR database.

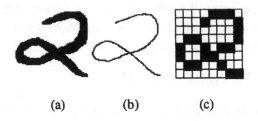

(a)   (b)   (c)

Figure 2. Thinning and normalization effects: (a) original bitmap,
(b) thinned image, (c) normalized image.

with the notation element/membership. We are entitled to ask where these fuzzy memberships come from and what they mean. To answer this question we use a voting model involving human voters. The world is not fuzzy. It is continuous and messy and we have to give labels to things we want to recognize as particular objects. There will always be borderline cases, and we can use graded membership in the nearest and most appropriate categories. Fuzziness arises when humans impose these categories – thus it is natural to use human judgment in developing the semantics of fuzzy sets.

Imagine that we have a representative set of people labeled 1 through to 10. Each person is asked to accept or reject the dice score of x as *small*. They can believe x is a borderline case, but they must make a binary decision to accept or reject. We take the membership of x in the fuzzy set *small* to be the proportion of persons who accept x as *small*.

Thus we know that everyone accepted 1 as small, 90% of persons accepted 2 as small and 30% of persons accepted 3 as small. We only know the proportions who accepted each score and not the complete voting pattern of each person. We will assume that anyone who accepted x as being small will also accept any score lower than x as being small. With this assumption we can write down the voting pattern:

1	2	3	4	5	6	7	8	9	10	persons
1	1	1	1	1	1	1	1	1	1	everyone accepts 1
2	2	2	2	2	2	2	2	2		90% accept 2
3	3	3								30% accept 3

Therefore, 1 person accepts $\{1\}$, 6 persons accept $\{1, 2\}$ and 3 persons accept $\{1, 2, 3\}$ as being the possible sets of scores when told the dice is small. If a member is drawn at random, then the probability distribution for the set of scores this person will accept is

$$\{1\} : 0.1, \{1, 2\}: 0.6, \{1, 2, 3\} : 0.3$$

This is a probability distribution on the power set of dice scores called a mass assignment and written as

$$m_{small} = \{1\} : 0.1, \{1, 2\}: 0.6, \{1, 2, 3\} : 0.3$$

It is, of course, a random set [7] and also a basic probability assignment of the Shafer Dempster theory [8]. We give it the name of mass assignment because we use it in a different way than Shafer Dempster theory. This mass assignment corresponds to a family of distributions on the set of dice scores. Each mass associated with a set of more than one element can be divided in some way among the elements of the set. This will lead to a distribution over the dice scores and there are an infinite number of ways in which this can be done.

Suppose we wish to give a unique distribution over the dice scores when we are told the dice value is small. How can we choose this distribution from the family of possible distributions arising from the mass assignment? To provide the least prejudiced (fairest) distribution, we would divide the masses among the elements of the set associated with them according to the prior for the dice scores. If this prior is unknown, then we can use a local entropy concept and divide each mass equally among the elements of its set. The resulting distribution is called the *least prejudiced distribution*. This corresponds to the probability of a dice score being chosen if we select a person from the voters at random and then ask that person to select one value from their possible set of accepted values when told the dice value is *small*.

For our case, when we know the dice value is *small* and assume the dice has a uniform prior, we obtain the least prejudiced distribution:

$$1 : 0.1 + 1/2(0.6) + 1/3(0.3) = 0.5$$

$$2 : 1/2(0.6) + 1/6(0.3) = 0.4$$

$$3 : 1/3(0.3) = 0.1$$

Thus,

$$\Pr(\text{dice is } 1 \mid \text{dice is } small) = 0.5,$$

$$\Pr(\text{dice is } 2 \mid \text{dice is } small) = 0.4,$$

$$\Pr(\text{dice is } 3 \mid \text{dice is } small) = 0.1$$

We will also use the notation

$$\text{lpd}_{small} = 1 : 0.5, 2 : 0.4, 3 : 0.1$$

where lpd stands for least prejudiced distribution.

We note that this is only a small modification from the following non-fuzzy case. If the dice score is even, then we can say that the scores 2, 4, and 6 are equally likely for a fair dice. In this case we put a mass of 1 with the set {2, 4, 6} and split this mass equally among the elements of the set of possible scores to get a least prejudiced distribution. The only difference between this and the fuzzy case is that the mass assignment comes from a crisp set rather than a fuzzy set.

3.2 Semantic Unification

Semantic unification enables us to calculate the probability that the dice value will be in a particular set, given that the dice value is known to be in some other set. In the crisp case we can ask for the probability that the dice value will be {1 or 2}, given that it is even.

$$\Pr(\{\text{dice value in } \{1, 2\} \mid \text{dice value is even}) = 1/3$$

Similarly, we can ask for the probability of the dice value being *about_2* when we know it is *small*, where *about_2* is a fuzzy set defined by

$about\_2 = 1 / 0.4 + 2 / 1 + 3 / 0.4$

For illustrative purposes, we will also assume that the dice is biased, with a prior

$1 : 0.1, 2 : 0.2, 3 : 0.3, 4 : 0.2, 5 : 0.1, 6 : 0.1$

This modifies the least prejudiced distribution for *small*

$1 : 0.1 + 1/3(0.6) + 1/6(0.3) = 0.35$

$2 : 2/3(0.6) + 2/6(0.3) = 0.5$

$3 : 3/6(0.3) = 0.15$

which is the distribution Pr(dice is i I dice is *small*).

The mass assignment for the fuzzy set *about_2* is

$m_{about\_2} = \{2\} : 0.6, \{1, 2, 3\} : 0.4$

We can use this mass assignment with the least prejudiced distribution for *small* to obtain a point value for Pr(dice value is *about_2* I dice value is *small*). From the least prejudiced distribution for *small* we obtain

$Pr(\{2\}|\,small) = 0.5, \ Pr(\{1, 2, 3\} \,|\,small) = 0.35 + 0.5 + 0.15 = 1$

and we define the Pr(dice value is *about_2* I dice value is *small*) as

Pr(dice value is *about_2* I dice value is *small*) =

$m_{about\_2}(\{2\})Pr(\{2\}|\,small) + m_{about\_2}(\{1, 2, 3\})Pr(\{1, 2, 3\} \,|\,small)$

$= 0.6 \cdot 0.5 + 0.4 \cdot 1 = 0.7$

We can see the equivalent of this in tabular form in Figure 3.

We are determining the probability of *about_2* when we are given *small*. So for each cell we wish to calculate the probability of the corresponding cell set associated with *about_2*, given the corresponding cell set associated with *small*. With each cell we associate a probability equal to the corresponding cell mass of *about_2* and the corresponding cell mass of *small*. Therefore, with the topmost cell we associate the probability (0.6)(0.1). We then determine the proportion of these cell probabilities which contribute to Pr(*about_2* | *small*).

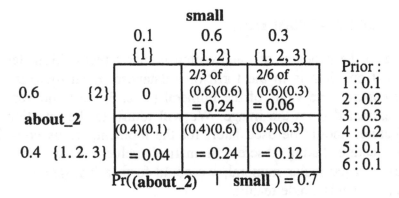

Figure 3. Semantic unification tableau for about_2, given small.

The left topmost cell has entry 0 since the truth of {1}, given {2}, is false. Thus for this entry, we take zero of the cell probability. The second entry in the top row is 0.24 since the truth of {2}, given 1, is false, but the truth of {2}, given 2, is true and 2 occurs with twice the probability of 1 according to the prior. Thus we take 2/3 of the cell probability. The third entry in the top row is calculated in a similar way. The leftmost entry in bottom row is 0.04, since {1, 2, 3} is true for {1}. Thus we take all of the cell probability, and do not have to split it. The other cells in the bottom row are similar - both {1, 2} and {1, 2, 3} are true, given {1, 2, 3}.

The use of the product rule to calculate cell probabilities assumes that the choice of *about_2* is independent of *small*. We did not take account of *small* when deciding to use *about_2*. Other semantic models can be defined. See [9], [10] for related work and the connection with random sets [7].

This process of determining Pr(*about_2* | *small*) is called point value semantic unification. We can also use an interval version, called interval semantic unification, where we do not use the prior to select between those values which give true and those values which give false. We simply record in each cell the probability, and label the cell with *t*   if the truth of the column set given the row set is true,

    f   if the truth of the column set given the row set is false,

    u   if the truth of the column set given the row set is uncertain.

3.3 Fuzzy Sets from Data

Returning to the problem of handwritten numeral recognition, let us assume that our data is in the form of a database. Each tuple in the database contains the value of the numeral (1, 2, 3, etc.) and various features extracted from the image. We will return to the question of which features should be used later; for the moment, let us consider just one feature taking discrete numerical values. Examining all examples of a particular numeral, we could form a histogram of the values taken by this single feature.

By assuming that this represents a least prejudiced distribution, we can form the corresponding fuzzy set and hence derive a rule of the form

> Classification of X is C
>   if
>   (feature$_1$ for X is F$_1$)

where F_1 is the fuzzy set corresponding to the feature values. The extension to the continuous case is simple and is outlined in [11], [12]. We now consider how to create more complex rules, involving several features.

3.4 Evidential Logic Rule of Combination

As in most pattern recognition problems, we are required to generalize from examples to new situations. A mechanism well suited to inductive reasoning approaches is the evidential logic rule [2], [3].

Evidential support logic provides a method of determining a match between known objects and similar unknown objects. The logic is case-based where features of the unknown object are partially matched with the features of a set of case example objects. The matching of features provides degrees of evidence in support of the match between objects, each feature being assigned an importance or weighting value (between 0 and 1), which reflects how much influence an individual feature has on the overall support for an object. The features themselves may be constructed from the conjunction, disjunction or negation of other sub-features through straightforward support logic calculus [13]. Features may also be defined by other rules.

(a)

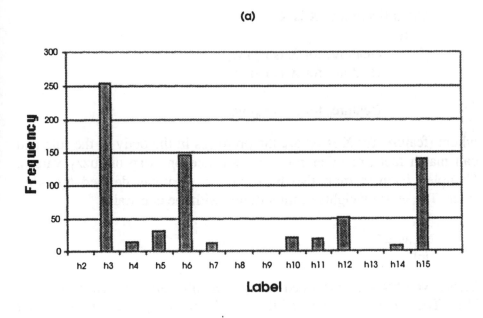

(b)

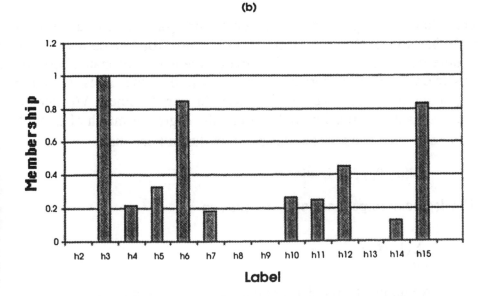

Figure 4. (a) Frequency count;   (b) Corresponding discrete fuzzy set.

Given a rule whose head contains a conclusion C to be inferred on the basis of n premises, evidential reasoning is used to determine the support for C as a function of the weighted sum of supports for the premises, as follows:

Classification of X is C
> if
>> (feature$_1$ for X is F$_1$) w$_1$
>> (feature$_2$ for X is F$_2$) w$_2$
>>
>> ---
>>
>> (feature$_n$ for X is F$_n$) w$_n$

where (feature$_i$ for X is F$_i$) are the premises in the body of the rule and can match facts, or other rules. F$_i$ can take on crisp or fuzzy values. Complex features may also be used, i.e., features defined by other rules. The w$_i$ are weights of importance with the constraint

$$\sum_{i=1}^{n} wi = 1.$$

These weights are calculated using semantic discrimination analysis [14]. They give a measure of the discriminating power the particular feature has with respect to a particular classification.

For rule k, the importance of a given feature (say F$_r$) depends on how well the associated fuzzy set, $\mathbf{f_{kr}}$, discriminates from the corresponding feature values in other rules. The smaller the degree of match between this fuzzy set and the corresponding ones in the other rules, the more important this feature becomes in discriminating the object. We can use point semantic unification to determine the degree of match of $\mathbf{f_{kr}}$ with $\mathbf{f_{ir}}$ for all i such that i≠k. Let this result in the unifications

$$\mathbf{f_{kr}|f_{ir}} : \theta_{kir} ; \text{ all i, i≠k.}$$

The degree of importance of F$_r$ depends on

$$\sum_{i \neq k}^{m} (1 - \theta_{kir})$$

so that we define the relative weights of features in rule k as

$$w'_{kr} = 1 - \frac{\sum_{\substack{i \neq k}}^{m} \theta_{kir}}{m - 1} \qquad \text{for } r = 1, \dots, n.$$

The relative set of weights for rule k, namely, $\{w_{k1}, \dots, w_{kn}\}$ are then normalized to give the importance weights for rule k

$$w_{ki} = \frac{w'_{ki}}{\sum_{j=1}^{n} w'_{kj}} \qquad \text{for } i = 1, \dots, n.$$

Each feature in the evidential rule body is matched to the features in the knowledge base and a degree of support is given for the feature, which is multiplied by the feature's importance. The overall support for the conclusion is given by the summation of these individual supports. The matching procedure is point semantic unification which provides a point value for the truth of **f | f'**.

4   Choice of Features

The three most popular explanations of pattern recognition are template, feature, and structural theories. A template theory proposes that people compare two patterns by measuring their degree of overlap. This fails to account for the great variability of patterns or variations in the position, orientation and size of presented patterns. Furthermore, templates do not allow for alternative descriptions of patterns. These problems usually lead us to dismiss template theories unless we are dealing with restricted patterns (e.g., printed characters). The most common theories of pattern recognition assume that patterns are analyzed into features. According to some feature theories, discrimination between patterns is achieved by discovering those features which are present in the stimuli and on which they actually differ. A distinctive feature of a pattern is any property on which it may differ from other patterns. Structural theories enrich feature theories by stating explicitly how the features of a pattern are joined together, thus providing a more complete description of a pattern.

The first step in both the feature and structural approaches is to consider which discriminatory features to select and how to extract these features. This initial choice is usually based on intuition and knowledge of the pattern generating mechanisms. The number of features needed to successfully perform a given recognition task depends on the discriminatory qualities of the chosen features. If all the initially proposed features are used in the pattern classification system, its performance may be inadequate, both in terms of misclassification rates and computation time. On the other hand, as the success or failure of a pattern recognition system is heavily dependent on the choice of good features (that separate the classes well), it is important not to limit *a priori* the number of features. Sometimes the choice of a new feature may yield good performance. Therefore, one should initially propose all the features that are potentially useful for the pattern classification system and use a feature selection procedure to choose a subset of those features according to quantitative criteria, ensuring that the feature subset is among the best that could be obtained from the original feature set. By finding appropriate combinations of features, the performance of the recognition system may be further enhanced.

4.1 Automatic Feature Discovery

Over the years, researchers have proposed a multitude of features (syntactic, topological, mathematical, etc.) that could be used in the recognition of handwritten characters [15]. Obviously such features have to be determined *a priori*, with the danger of overlooking important ones. Moreover, the extraction of such features usually requires complex image processing algorithms. We allow our system to automatically discover features directly from the bitmaps.

We split the normalized bitmaps of the handwritten digits into 4 distinct regions: A, B, C, D as shown in Figure 5. After image normalization, we identified approximately 8000 different bitmap patterns that could occur in any of the 4 regions. We could define fuzzy sets on the bitmaps but that would be computationally expensive. In addition, it would be impossible to assess intuitively the validity of the resulting features. Instead, we opted to reduce the dimensionality of the data by mapping the different bitmaps onto different categories of lines and curves. These were automatically defined by a Kohonen self-organizing map (SOM). The SOM uses unsupervised learning to modify the internal state of the network to model the features found in the training data. This SOM is a special type of competitive learning network that defines a spatial neighborhood for each output unit. The node vector to which inputs match is selectively optimized to represent an average of the training data. Then all the training data are represented by the node vectors of the map. The obvious problem is deciding on the optimum number of output units such a network should have. Too many features are likely to reduce computational efficiency while, on the other hand, mapping the input patterns on too few output nodes may result in loss of information and compromise the generalization properties of the features.

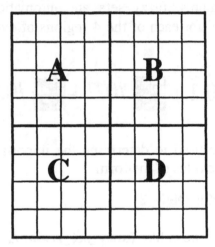

Figure 5. The 8 × 8 bitmap is split into 4 regions: A, B, C and D.

There are a number of models that dynamically change structure and form of the SOM during training. We have adapted Fritzke's [16] proposal of an incremental self-organizing network that is able to automatically determine a height/width ratio suitable for the data at hand. The advantages of having a rectangular map are that the network can be trivially displayed by drawing a grid and that the implementation does not require the handling of graphs or other sophisticated data structures; an array suffices to represent the network.

Training occurs in two stages. The purpose of the first stage is to find the most suitable dimensions for the SOM. Starting with a 2 × 2 topology, where weights are initialized to random values, new rows and columns are inserted as long as the network has less than 30 units. During this phase, the neighborhood and learning rate are held constant. The process is repeated a number of times and we look for convergence in the different maps. The different results should converge to similar mappings with the same dimensions. The converging topology is used for the final mapping of bitmaps onto generic lines and curves. In the second stage, training proceeds with the fine-tuning of the output vector positions. Neighborhoods are reorganized with the learning rate decreasing linearly with time. By converting the weights in the network to grayscale values we can visualize the resulting nodes as representations of the curves and lines the SOM has produced, as shown below.

Based on these features, fuzzy sets are automatically generated to produce a 'signature' for each of the 4 regions of every digit, as in the following example:

$$\mathbf{f} = \blacksquare /0.1 + \blacksquare /0.7 + \blacksquare /0.75 + \blacksquare /1$$

These fuzzy sets are used to determine the importance weights. This gives evidential logic rules of the form

((class of character X is **Digit_2**)
    (evlog (
        (region a of X looks like **fuzzy_set_regionA_digit2**)   0.20241
        (region b of X looks like **fuzzy_set_regionB_digit2**)   0.24598
        (region c of X looks like **fuzzy_set_regionC_digit2**)   0.36402
        (region d of X looks like **fuzzy_set_regionD_digit2**)   0.18759
)))

4.2 Compound Features

In Section 3, we discussed how single attributes are used as features F_i in the evidential logic rule. Discovering structural relationships in the original features F_i, to produce new compound features, may help us describe the training set more effectively and concisely while, at the same time, we are likely to obtain better generalization on the unseen test set. Compound features can be made up of a conjunction of features, as in "feature A has value X AND feature B has value Y in all instances of class C."

We will now identify how the fuzzy sets associated with such compound features can be formed. If the feature F is a combination of two attributes, say X_i and X_j, then the part of the body of the rule corresponding to this attribute will take the form

F is **f**   where

$$\mathbf{f} = \quad h_{i1} \wedge h_{j1}/\mu_{11} + h_{i1} \wedge h_{j2}/\mu_{12} + \ldots + h_{i1} \wedge h_{jn}/\mu_{1n}$$

$$h_{i2} \wedge h_{j1}/\mu_{21} + h_{i2} \wedge h_{j2}/\mu_{22} + \ldots + h_{i2} \wedge h_{jn}/\mu_{2n}$$

$$\ldots$$

$$h_{im} \wedge h_{j1}/\mu_{m1} + h_{im} \wedge h_{j2}/\mu_{m2} + \ldots + h_{im} \wedge h_{jn}/\mu_{mn}$$

where $\{h_{ik}\}$ for $k=1,\ldots,n$ is the set of values that X_i can take and $\{h_{jk}\}$ for $k=1,\ldots,m$ the set of values on X_j.

The fuzzy set **f** is a fuzzy set on the product space of $\{h_{ik}\}$ and $\{h_{jk}\}$, namely $\{h_{ik}\} \times \{h_{jk}\}$.

For our problem, we will consider the following feature combinations: A, B, C, D, $A \wedge B$, $A \wedge C$, $A \wedge D$, $B \wedge C$, $B \wedge D$, $C \wedge D$, $A \wedge B \wedge C$, $A \wedge B \wedge D$, $A \wedge C \wedge D$, $B \wedge C \wedge D$. We ignore $A \wedge B \wedge C \wedge D$ since this reproduces only the training pattern and, as such, provides no generalization to unseen patterns. The domain of A is, as before, the set of all the nodes in the SOM, while the domain for $A \wedge B$, for example, is the Cartesian product space of A and B. For X features in the pattern, we get $2^X - 1$ possible combinations and if the feature can take N different values, the fuzzy set will have $N^{|X|}$ elements. This is only the theoretically worst case. In practice, the possible combinations and the number of elements in the fuzzy set are determined by the actual training set. Furthermore, we can improve the computational efficiency of the evidential logic rule by ignoring features with low weights and renormalizing the remaining weights.

5   Implementation and Results

We have implemented the system in the logic programming language FRIL [13], [17], [18]. The Fril environment provides the support logic rules of inference, the evidential logic rule, semantic unification and built-in facilities to represent and reason with fuzzy sets and probabilistic uncertainties. The resulting architecture incorporates distinct modules for learning, processing and classification. The system can be rather slow to train (up to 10 hours on an average workstation depending on whether or not compound features are used) but produces a fast classifier managing 5 to 15 digits per second without special efforts to optimize the code. In the classification stage, memory requirements are negligible.

The classifier produced a support for each pattern in the database being recognized as each of the 10 digits. We see an example in Figure 6.

To obtain a recognition rate, we chose to assign input patterns to the class with the highest support. Results on both the training and test sets are shown in Table 2. The table gives the percentage of numerals for which the highest output was associated with the true class (the recognition rate) which is a standard measure of performance for character recognition systems. A moderate recognition rate of 73.2% is achieved on the training set, and only 60.2% on the test set. This strongly indicates that the automatically discovered features are not the best possible, especially with respect to their generalization properties.

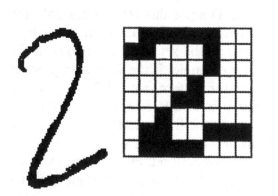

digit	support
1	0.08
2	**0.99**
3	0.78
4	0
5	0.09
6	0
7	0.05
8	0.26
9	0.06

Figure 6. Sample of test results showing original image (left),
thinned and normalised image (centre) and classification (right).

In an attempt to improve the poor results obtained with the features discovered by the SOM, we used structural combinations of these features as discussed in Section 4.2. Table 3 lists the results obtained with the rules formed on all 14 features (simple and compound).

Table 2. Results for Simple Features.

DataSet	Class										
	0	1	2	3	4	5	6	7	8	9	Total
Training	66.2	79.8	70.4	69.2	81.2	71.5	66.3	63.4	77.4	82.2	73.2
Testing	51.0	67.6	55.9	51.4	82.1	39.7	68.1	54.1	45.6	49.6	60.2

Table 3. Results for all Simple and Combined Features.

DataSet	Class										
	0	1	2	3	4	5	6	7	8	9	Total
Training	80.3	83.2	90.9	75.4	94.6	89.5	83.0	66.5	93.3	91.4	84.6
Testing	73.7	77.6	80.8	61.4	92.5	49.1	78.1	64.2	55.6	59.7	71.9

Using structural combinations of the features improves the results considerably to 84.6% and 71.9% on the training and test sets, respectively. The computational overhead incurred by using rules with more features and bigger fuzzy sets is significant, but can be reduced by discarding features that have low weights. This can be done simply by imposing an arbitrary threshold on the minimum weight a feature should have, e.g., for the 14 features

$$\frac{1/14}{2} = 0.036$$

We then have to renormalize the weights of the remaining features. This reduces the recognition accuracy to 82.3% on the training set and 70.8% on the test set, but more than doubles processing speed.

As can be seen in Table 4, combinations of only 3 features yield even better results. These are the best results (89.6% on the training set and 82.5% on the test set) we have obtained on this dataset with the features produced by the SOM. We get an improvement in the results (compared to those in Table 3), because combinations of three features seem to provide the right balance between generalization and proximity to the original patterns.

Table 4. Results for Combinations of 3 Features.

DataSet	Class										
	0	1	2	3	4	5	6	7	8	9	Total
Training	88.6	85.3	96.2	79.0	98.3	88.6	86.6	83.2	98.7	95.4	*89.6*
Testing	81.6	82.6	85.8	65.5	97.5	69.1	84.1	70.1	61.6	61.7	*82.5*

Some other systems used for the same task claim recognition accuracy in excess of 90% on the test set. However, these are usually achieved with carefully chosen initial features that require complex image processing techniques (e.g., 97% with Kirsch differential edge detectors in [19], 94% with complex Fourier descriptors from the outer contours of numerals and simple topological features from the inner contours in [20]), 96.5% with generative models in [21] and 97.9% with radial basis function networks in [22]). The relative merit of our approach lies in the *automatic* discovery of features and the significant improvement that *structural combinations* of these features effect on correct classification rates. Furthermore, it is now generally accepted that high-performance recognition of handwritten characters will be achieved only by the use of context. To exploit contextual knowledge, several most likely classifications must be considered for each character. This criterion is satisfied by our system in that the classifier produces a membership for each digit belonging to each class. A closer inspection of our results revealed that 96.5% of the misclassified digits were assigned to the correct class with the second highest membership value.

6   Summary

We have described a methodology for automatic discovery and structural combination of discriminating features from data so that patterns can be effectively recognized. Our methodology depends on a self-constructing self-organizing network as a preprocessor which clusters the original data in an optimal number of output classes. Statistical data and mass assignments are subsequently used for intelligent feature extraction from example data. Such features, as well as combinations of these features, are used in the evidential logic rule which infers the truth of classifier rules based on the truth of the premises. The importance of the features is determined through the use of semantic discrimination analysis.

The system was tested on a database of handwritten digits. Initial results have been positive and demonstrate well the significant improvement that structural combination of features effect on the recognition rate.

Future work that will further improve the impact of the structural features on classification accuracy is to use all logic operators. The problem that arises is deciding which operators to use and how to combine the features to construct higher order features that optimally discriminate. Genetic programming [23], [24] can be used to perform a multipoint search through the possibilities and determine the most appropriate combinations that capture structural aspects of the patterns. To achieve this, the discrimination power of the features can be used to guide the search (i.e., as part of the fitness function).

Furthermore, rather than splitting the bitmaps into four regions of equal size, different sized windows could be passed over the character to discover features and use the relative position and size of these windows to produce structural features. The major difficulty with this approach is the huge number of possible window sizes and positions that would have to be examined.

Finally, the variation in the recognition rates of individual digits suggests that the bitmaps could be more effectively clustered according to the features discovered by the SOM. Categorization will no longer depend purely on the class labels given in the training set, but will also take into account the semantic similarity of the objects. A second classifier can also be devised, more or less dedicated to the mistakes and marginal decisions of the original classifier. This is a procedure that can be iterated to produce a hierarchical classifier that yields further improvements in recognition accuracy.

Acknowledgments

This work has been partially supported by British Airways plc. under EPSRC CASE award 9356105X and the Leventis Foundation, Greece.

References

[1] Zadeh, L.A. (1997), "The Roles of Fuzzy Logic and Soft Computing in the Conception, Design, and Deployment of Intelligent Systems," in *Software Agents and Soft Computing: Concepts and Applications*, H.S. Nwana and N. Azarmi, Editors. 1997, Springer (LNAI 1198), pp. 183-190.

[2] Baldwin, J.F. (1987), "Evidential Support Logic Programming," *Fuzzy Sets and Systems*, Vol. 24, pp.1-26.

[3] Baldwin, J.F. (1993), "Evidential Reasoning, Fril, and Case-Based Reasoning," *International Journal of Intelligent Systems*, Vol. 8, pp. 939-961.

[4] Kohonen, T. (1984), *Self -Organisation and Associative Memory*, Springer.

[5] Hull, J.J. (1994), "A Database for Handwritten Text Recognition Research," *IEEE Transactions on Pattern Analysis and Machine Intelligence*, Vol. 16, pp. 550-554.

[6] Stentiford, F.W.M. and Mortimer, R.G. (1983), "Some New Heuristics for Thinning Binary Handprinted Characters for OCR," *IEEE Transactions on Systems, Man and Cybernetics*, Vol. 13, pp. 81-84.

[7] Nguyen, H.T. (1978), "On Random Sets and Belief Functions," *J. Math. Anal. & Appl.*, Vol. 65, pp. 531-542.

[8] Shafer, G. (1976), *A Mathematical Theory of Evidence*, Princeton University Press.

[9] Dubois, D. and Prade, H. (1982), "On Several Representations of an Uncertain Body of Evidence," in *Fuzzy Information and Decision Processes*, M.M. Gupta and E. Sanchez, Editors, North Holland.

[10] Dubois, D. and Prade, H. (1991), "Random Sets and Fuzzy Interval Analysis," *Fuzzy Sets and Systems*, Vol. 42, pp. 87-101.

[11] Baldwin, J.F. (1993), "Fuzzy Sets, Fuzzy Clustering, and Fuzzy Rules in AI," in *Fuzzy Logic in AI*, A.L. Ralescu, Editor, Springer Verlag. pp. 10-23.

[12] Baldwin, J.F. and Martin, T.P. (1997), "Basic Concepts of a Fuzzy Logic Data Browser with Applications," in *Software Agents and Soft Computing: Concepts and Applications*, H.S. Nwana and N. Azarmi, Editors, Springer (LNAI 1198). pp. 211-241.

[13] Baldwin, J.F. (1986), "Support Logic Programming" in *Fuzzy Sets - Theory and Applications*, A. Jones, Editor, D. Reidel, pp. 133-170.

[14] Baldwin, J.F. (1994) "Evidential Logic Rules from Examples," in *EUFIT-94*. Aachen, Germany.

[15] Trier, O.D., Jain, A.K., and Taxt, T. (1986), "Feature Extraction Methods for Character Recognition – a Survey," *Pattern Recognition*, Vol. 29, pp. 641-662.

[16] Fritzke, B. (1995), "Growing Grid - a self-organizing network with constant neighborhood range and adaptation strength," *Neural Processing Letters*, Vol. 2, pp. 1-5.

[17] Baldwin, J.F., Martin, T.P., and Pilsworth, B.W. (1988), *FRIL Manual (Version 4.0)*, Fril Systems Ltd., Bristol Business Centre, Maggs House, Queens Road, Bristol, BS8 1QX, U.K. pp. 1-697.

[18] Baldwin, J.F., Martin, T.P. and Pilsworth, B.W. (1995), *FRIL - Fuzzy and Evidential Reasoning in AI*, U.K.: Research Studies Press (John Wiley).

[19] Lee, S.W. (1996), "Off-Line Recognition of Totally Unconstrained Handwritten Numerals Using Multilayer Cluster Neural Networks," *IEEE Trans. on Pattern Analysis and Machine Intelligence*, Vol. 18, pp. 648-652.

[20] Krzyzak, A., Dai, W., and Suen, C.Y. (1990). "Unconstrained Handwritten Character Classification Using Modified Backpropagation Model," in *First Int. Workshop on Frontiers in Handwriting Recognition*. Montreal, Canada.

[21] Revow, M., Williams, C.K.I., and Hinton, G.E. (1996), "Using Generative Models for Handwritten Digit Recognition," *IEEE Transactions on Pattern Analysis and Machine Intelligence*, Vol. 18, pp. 592-606.

[22] Lemarie, B (1993). "Practical Implementation of a Radial Basis Function Network for Handwritten Digit Recognition," in *Second Int. Conf. on Document Analysis and Recognition*. Tsukuba, Japan.

[23] Koza, J.R. (1992), *Genetic Programming*, MIT Press.

[24] Holland, J.H. (1975), *Adaptation in Natural and Artificial Systems*, University of Michigan Press.

Chapter 5:

Handwritten Character Recognition Using a MLP

HANDWRITTEN CHARACTER RECOGNITION USING A MLP

F. Sorbello, G.A.M. Gioiello and **S. Vitabile**

Department of Electrical Engineering - University of Palermo

Viale delle Scienze, 90128, Palermo, Italy

In this chapter we describe some algorithms and techniques intended for the handwritten character recognition task using a MultiLayer Perceptron neural network. The adopted strategies, in the learning and in the testing phase, lead to high performances in terms of recognition rate, preprocessing and classification speed. We tested the proposed approach using the National Institute for Standard and Technology handwritten character database. The simplicity of the chosen techniques allows either an implementation on a traditional serial computer still having reasonable performances or a digital implementation of the MLP neural network. The obtained experimental results are also reported.

1 Introduction

The handwritten character recognition problem still constitutes an open research field for the definition of a successful strategy combining high recognition rate with favorable speed.

Several approaches have been proposed by researchers to solve in an optimal way the handwritten character recognition task including statistical solutions [1], neural solutions [2], [3], and hybrid systems combining neural with statistical aspects [4]. Even if all these approaches are valid, we must still evaluate them using a set of parameters such as the recognition and speed rate, memory requirements, possibility of a digital implementation and, for neural solutions, the learning speed. On this basis, the neural solution is more attractive because it offers a very good compromise between recognition rate, recognition speed and memory/hardware resources. In particular, MultiLayer Perceptrons (MLP) are widely used for classification problems [6]. The capability to approximate boundary surfaces of arbitrary complexity makes the MLP classifiers universal [7].

In this chapter, we present a methodology based on a set of appropriate choices for the handwritten character recognition task using a MLP. The use of simple preprocessing techniques, the enhancement of the training set, the choice of opportune activation functions combined with a two step learning process, a learning procedure based on Powell's conjugate gradient optimization algorithm [8], and a triple presentation of the test samples are the main ingredients of a recipe that yields results very interesting in terms of both recognition rate and speed.

In what follows we will present a full description of each of these techniques and the experimental results obtained by the software and hardware system implementation. Either the full software (C-code) implementation of the preprocessing handwritten characters technique and the MLP neural network or the digital implementation of the proposed neural architecture, using the VHDL language, gives high recognition rate and speed on a traditional serial computer. Training and testing of the MLP were performed using the whole National Institute for Standard and Technology (NIST) database.

2   The Adopted MLP Description

The *perceptron* was introduced by F. Rosenblatt [5] in 1959 as a way to use error-correction to train a neural network to respond to a specific set of patterns.

It is a feed-forward neural network with n inputs, connected to other perceptrons or to the outside world, and a single output. Each input x_i is weighted, i.e., multiplied by some synaptic weight w_i. The weighted inputs are then summed together, a bias term is added, and a nonlinear transformation is accomplished on the sum. The nonlinear transformation is analogous to the sigmoid relationship between the excitation and the frequency of firing observed in the biological neurons. Consequently, the output of the perceptron neuron can be seen as

$$z = f_h (\sum_{i=0}^{n-1} w_i x_i + \theta) \tag{1}$$

where f_h is the nonlinear transformation function, w_i are the synaptic weights, θ is the neuron bias term, and x_i the previous inputs from other perceptrons or from the outside world. Training the perceptron to solve a pattern recognition problem amounts to arranging the weights such that in the region of one class of input objects, the output of the perceptron would be high. When the features represent the other class of input objects, the output of the perceptron would be low. So the perceptron learning procedure can be seen as an automatic method of setting the weights using only the training data.

The single-layer perceptron had the limitation that it can discriminate only between classes that are hyperplane separable. The multilayer artificial neural networks can be used for interesting problems requiring greater discrimination. In these networks the learning algorithm also allows the training of the weights on the hidden layers in order to allow the formation of complex decision regions in the feature space. Thus, the MultiLayer Perceptrons can be used to solve classification problems with nonlinearly separable input classes. In particular, an existence proof shows that three layers are sufficient for any given classification problem [22].

Several forms of nonlinear transformation functions can be used as well as different learning algorithms in order to find the correct set of weights for the neural network. In the experiments described in this chapter, we adopt a two-layer perceptron with the sinusoidal activation function for the hidden layer and the linear activation function for the output layer. The learning algorithm used for adapting the neural weights is the Conjugate Gradient Descent optimization algorithm proposed by Powell [8]. In what follows we describe the characteristics of both the used learning algorithm and the used activation functions.

2.1 The Learning Algorithm

Several experiments using different topologies and learning-testing paradigms were conducted. In all cases the Conjugate Gradient Descent optimization algorithm proposed by Powell [8] has been used since it provides better results. Particular features of this algorithm are

- high learning speed;
- optimal performances for high-grade polynomial functions;

- reasonable amount of memory needed: memory requirements increase linearly with the number of variables.

The chosen learning algorithm is faster than the steepest descent algorithm and does not require any choice of critical parameters like momentum or learning rate. The search of the minimum of the error surface is based on the determination of a set of directions, so that the minimization gained on a particular direction does not interfere with the minimization obtained before. The algorithm employs the optimization technique of the conjugate gradient descent algorithm, borrowed from the numerical analysis field. Such a technique substitutes the steepest descent method with one of minimization along a set of noninterfering directions. These directions are usually called *conjugate directions*. We can start by noting that minimizing a function f along a direction $\mathbf{u}$ brings us to a point were the gradient of f is orthogonal to $\mathbf{u}$, (otherwise, the minimization along $\mathbf{u}$ is not complete). Assume a point $\mathbf{P}$ as origin of the related associated coordinate system. A generic function f may be approximated near $\mathbf{P}$ by its Taylor series

$$f(x) = f(P) + \sum_i \frac{\delta f}{\delta x_i}\bigg|_P dx_i + \frac{1}{2}\sum_i \frac{\delta^2 f}{\delta x_i x_j}\bigg|_P dx_i dx_j + \dots$$

$$\approx c - b \cdot x + \frac{1}{2}x \cdot A \cdot x \tag{2}$$

where

$$c \equiv f(P), \quad b = -\nabla f\big|_P, \quad [A]_{ij} \equiv \frac{\partial^2 f}{\partial x_i \partial x_j}\bigg|_P \tag{3}$$

The $\mathbf{A}$ matrix whose components are the second-order partial derivatives of the f function is called the Hessian matrix of the function f in the point $\mathbf{P}$. When Equation (2) holds, the gradient of f is calculated as

$$\nabla f = A \cdot x - b \tag{4}$$

The gradient changes when moving along a direction as stated in Equation (5)

$$\delta(\nabla f) = A \cdot (\delta x) \tag{5}$$

Suppose we moved along direction **u** toward a minimum. The condition that motion along a new direction **v** does not spoil the minimization along **u** is simply that the gradient stays perpendicular to **u**. By Equation (5) this condition is expressed as

$$0 = u \cdot \delta(\nabla f) = u \cdot A \cdot v \qquad (6)$$

When Equation (6) holds for both the **u** and **v** directions, they are called conjugate directions. If Equation (6) is completely satisfied by a set of directions, this set is called a *conjugate set*.

An optimal result might be to have a set of N linearly independent conjugate directions. In this case only a single minimization step along these N directions is necessary to bring the desired minimum for a quadratic form as shown in Equation (2). If the form is not quadratic, the above described steps must be iterated in order to obtain a quadratic convergence toward the minimum.

The above-mentioned N mutually conjugate directions can be automatically searched with a recursive formula, developed by Powell [8].

2.2 The Activation Functions

The choice of the activation functions for the hidden and output layers is very critical to reach a good classification rate in each task. Several studies have been carried out on the choice of functions that lead to a better global minimum of the error function, searching to escape from unsatisfactory local minima.

Gori and Tesi in their work [12] affirm that by using linear functions the backpropagation algorithm reaches the global minimum of the error function if the classes are linearly separable. However, the choice of linear functions opposes the proof that a neuron transformation is based on a nonlinear function. Allen and Stork [13] have postulated the constraint that the activation function grows monotonically. In this direction, they carried out a study on the choice of the activation function that leads toward good results in terms of classification.

The criterion we followed to choose the activation function for the hidden layer starts from the examination of the XOR problem. It is well known that this problem cannot be classified by a perceptron with only one hidden unit with a sigmoidal activation function or with any growing monotonic function.

Such a function $f(I_1 \cdot w_1 + I_2 \cdot w_2 + \theta)$ would not classify the input points $(I_1 = 1, I_2 = 0)$ and $(I_1 = 0, I_2 = 1)$ with a high output value and the input $(I_1 = 1, I_2 = 1)$ with a low output value. If we consider a sinusoidal function, the above-mentioned conditions can be satisfied thanks to the periodicity of this function. Moreover, it is easy to prove that a sinusoidal function separates the classes by parallel straight lines and, in the case of multidimensional inputs, by parallel hyperplanes. For the XOR problem, the general expression of the activation function will be

$$O(\overline{I}, \overline{w}) = \sin(I_1 \cdot w_1 + I_2 \cdot w_2 + \theta) \tag{7}$$

The separation boundary between the two classes is obtained from Equation (7) by imposing the condition

$$I_1 \cdot w_1 + I_2 \cdot w_2 + \theta = k\pi \qquad k = \pm 1, \pm 2, \ldots \tag{8}$$

and this represents, on the input space, the equation of parallel straight lines with angular coefficient $-w_1 / w_2$ (see Figure 1). Therefore, the use of a sinusoidal activation function for the hidden layer guarantees a better separation of the classes, limiting the number of hidden units. To solve the XOR problem and N-bit parity problem just one hidden unit is necessary.

The experimental trials confirm this assertion, showing a better recognition rate for handwritten characters when a sinusoidal rather than sigmoidal activation function is used. This is also due to the shape of a sinusoidal function that does not have large saturation areas for which the gradient is low, as those with a sigmoidal function. Another advantage may arise if we want to realize a digital implementation of the network. It is very simple to implement a sinusoidal function in an accumulator, as described in a following section.

Equally important for the classification is the choice of the activation function for the output layer. Several authors [12] [14] [15] have shown the relation between the output activation function and the presence of local minima in the error surface. Gorse et al. [14] have proved experimentally that the probability of achieving a global minimum is higher when a linear rather than a sigmoid activation function is employed.

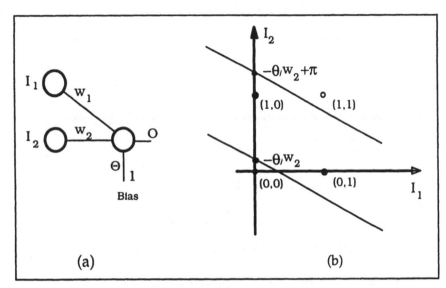

Figure 1. (a) Perceptron employed for the XOR problem;
(b) separation boundary of the two classes obtained using the sine function.

The trials performed showed that the learning speed and the recognition rate are slower with a sigmoidal function in the output layer (the latter by about two percentage points) if compared to the values obtained with the linear function. This is due to the particular separation made by a linear function that better takes into account the distribution of the elements in each class rather than the distribution of the boundary elements.

Let us consider two monodimensional classes: the * class and the o class. With the distribution of the elements shown in Figure 2, the threshold of the sigmoidal function σ_0 will separate the two classes at some point between pattern 5 of the * class and pattern 1 of the o class. The sigmoid threshold is greatly influenced by the boundary pattern 5 of the * class, because the sigmoid minimizes the energy function associated with the network.

Using the linear function, we include the possibility that pattern 5 of the * class may be an equivocal point and, consequently, should not constrain the determination of the separation line. In fact, if the linear function is a straight line of small slope s, the threshold value η_0, separating the two classes, coincides with the baricenter of the pattern's distribution. Therefore, η_0 will be closer to the * class elements than the sigmoid threshold σ_0.

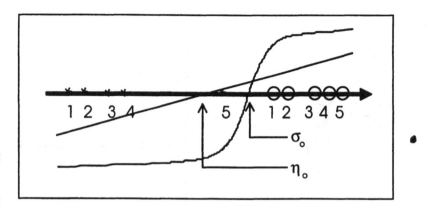

Figure 2. Discrimination thresholds for the sigmoidal and linear activation functions.

To sum up, the linear function is applied to the presynaptic values of each network's output; the first class represents the set of elements recognized by that output as belonging to the associated character class, and the second class represents the set of the other character elements.

The particular technique employed during the training of the MLP for the handwritten character recognition task uses the function shown in Figure 3. This is a piecewise linear function. If the input pattern belongs to the examining output unit class, the output will be cut off at the μ_{sat} value if the presynaptic activation σ is greater than a threshold value τ; otherwise the output will be proportional to the presynaptic activation σ.

The training of the neural network is carried out using the function described above with a slope $\sigma_1 = 0.5$ until a recognition rate of about 98% on the training set is achieved. Successive learning is made with a function slope $\sigma_2 < \sigma_1$ achieving a recognition rate of 100% on the training set. The final 100% score obtained after the training phase on the same training set is to be referred to the label proposed by the NIST. In the trials made, a $\sigma_2 = 0.1$ value has been chosen. This technique distances the patterns in output space so it is easier to separate them.

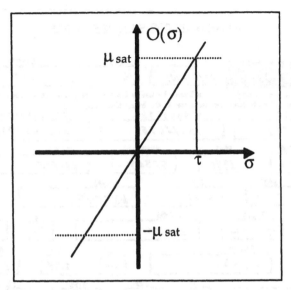

Figure 3. Activation function employed for the output layer.

To sum up, the experiments show that the sigmoid function leads to a slower learning speed and a greater tendency to get stuck in local minima, if compared to the sinusoidal function for the hidden layer and to the linear function for the output layer.

3 The Handwritten Character Recognition Task

We can decompose the character recognition task into a set of different phases. Each phase plays a determining role in obtaining a good classification rate. The main phases we distinguish are the following:

- data acquisition;
- character segmentation;
- character preprocessing;
- character processing.

The first two steps are bypassed in this study since we use the handwritten characters database supplied by NIST. The NIST database contains several forms with handwritten uppercase and digit characters. An example of the available data form is shown in Figure 4. Our efforts were aimed to optimize the latter two phases of the recognition process by performing a set of simple but effective techniques.

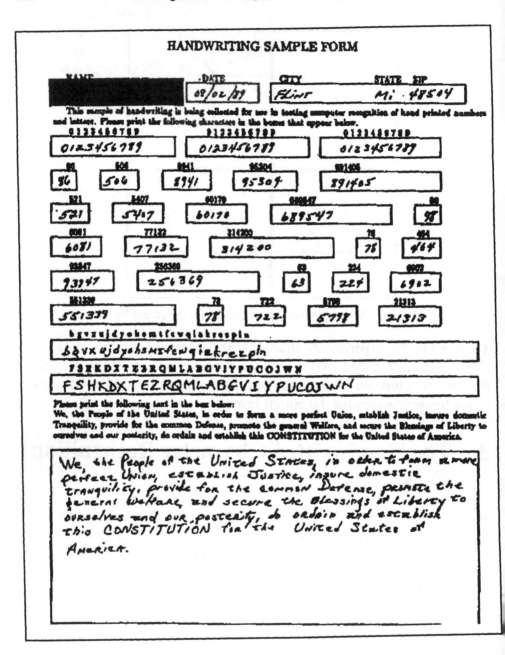

Figure 4. An example of the available NIST handwritten forms.

The proposed solution is a neural one. We adopted a MLP with one hidden layer having the sinusoidal function as hidden layer activation function and the linear function as output layer activation function (see Section 2.2). The adopted MLP was trained off-line using the conjugate gradient descent optimization algorithm proposed by Powell (see Section 2.1). In what follows is presented the description of the employed techniques for the latter two phases of the recognition process along with a theoretical justification.

3.1 The Preprocessing Phase

Printed characters have a defined writing direction (inertial axis centered in respect to the bitmap) and an established letterpress style (therefore, a fixed size). On the contrary, handwritten characters can be oriented in any direction on the plane and their size may not be constant. Therefore, it is essential to define an efficient strategy in order to reduce the classification error as much as possible. In the handwritten character recognition task, classification rate is mainly affected by characters' rotation and thickness or characters' size variation.

These characteristics may be obtained using an effective preprocessing technique normalizing each character. The starting characters consist of 128*128 bitmaps. The first operation carried out in the preprocessing stage consists of the matrix dimension reduction from 128*128 to 24*24 pixels. This reduction process consists of the following steps:

1. determine a rectangle tangent to the processed character, with sides parallel to the vertical and horizontal axes;
2. find the coordinate values x_{max}, y_{max}, x_{min}, y_{min} of the rectangle vertices;
3. calculate the (x',y') coordinates for a generic point of the new 24*24 matrix corresponding to the (x,y) coordinates of the black pixels of the original 128*128 matrix by using the following relations with $n = 24$:

$$x' = \frac{x - x_{min}}{x_{max} - x_{min}} * n \qquad y' = \frac{y - y_{min}}{y_{max} - y_{min}} * n \qquad (9)$$

If the black character in the 128*128 frame is smaller than 24 pixels (width or height), the transformation introduces white lines. This improves the recognition phase.

$$X = \begin{bmatrix} 0 & 1 & 1 & 1 & 1 & 1 \\ 0 & 0 & 0 & 0 & 1 & 1 \\ 1 & 0 & 0 & 0 & 1 & 1 \\ 0 & 0 & 0 & 0 & 1 & 0 \\ 0 & 0 & 0 & 1 & 1 & 0 \\ 0 & 0 & 0 & 1 & 1 & 0 \end{bmatrix} \qquad M = \begin{bmatrix} 3 & 7 \\ 0 & 5 \end{bmatrix}$$

Figure 5. An application of Equation (10) for a starting matrix 6*6.

A second reduction is then made by subdividing the 24*24 matrix into submatrices of 3*3 pixels and codifying each of these with an integer number included in the range [0,9] which represents the number of black pixels in the 3*3 matrix.

The new 8*8 matrix is obtained by applying the following equation:

$$M(i, j) = \sum_{n=3(j-1)+1}^{3(j-1)+3} \left[\sum_{k=3(i-1)+1}^{3(i-1)+3} X(k, n) \right] \qquad (10)$$

where X is the original 24*24 matrix and M is the new reduced 8*8 matrix. Therefore, the M matrix is composed of integers ranging from 0 to 9, while the elements of the X matrix are the integers 0 and 1. In this way it is possible to partially preserve information from the original matrix. Figure 5 presents an example of this technique for a starting matrix 6*6, obtaining a 2*2 matrix.

3.2 The Character Processing Phase

3.2.1 The Training Phase

The training (also called learning) phase plays an important role in the performance of a neural network. It is important to choose an appropriate training set, activation function and learning algorithm.

One of the more attractive aspects of a neural network is its generalization capability. The network acquires it by the learning process and the generalization degree achieved is related to the training set characteristics. The network will show good generalization capabilities if the training set is constituted of a large number of elements compared with the number of classes and equally distributed with respect to each class [2].

In the handwritten character recognition task, the neural network is also required to classify slightly rotated characters. A number of techniques are reported in the literature for rotated characters. These techniques require *a priori* knowledge in order to give the invariance property to the network with respect to slight rotations applied to the characters. The main techniques used in this direction may be classified as follows [9]:

- invariance by using appropriate network structures;
- invariance by using an appropriate training;
- invariance by constraint on the network's output trajectories described by the rotated input patterns.

The last technique is very difficult to apply because it requires either *a priori* knowledge of the functions trajectory associated with the transformation or the research of a set of parameters identifying invariance transformation.

Invariance by structure [10] is too onerous because it requires the evaluation of the mathematical relation between the weights of the input-hidden layer link due to the invariance transformation. Significant results are achieved by Le Cun et al. [11]. In this study they demonstrated that the recognition error may be reduced by keeping the gradient of the output functions low along the curve corresponding to the rotation of the training set patterns. Drawbacks of this method include the mathematical evaluation of the tangent vector at the curve representing the transformation on the input space and the restricted training set size (320 elements), which is small compared with the NIST database dimension.

A simple and efficient method to keep the gradient low consists of an artificial enhancement of the training set by the addition of slightly rotated characters. This technique leads to the smaller gradient values of the output function (associated at any neuron of the output layer) along the curve bound to the transformation. The addition of training samples for each class and the output functions associated to the network are constrained to cross for more points along the curve with the same target, rather than for just one point, thus limiting the gradient of this function around that pattern.

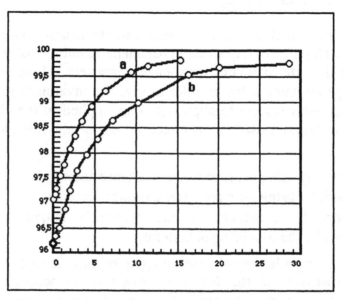

Figure 6. Rejection curve obtained with the uppercase (a) and digits (b) characters of NIST test set.

Let $T(\alpha,u)$ be the transformation that rotates the element of the training set u by an arbitrary angle α. We add two characters obtained by applying the rotation T with angle values of about ±0.3 rad for each NIST database character, decreeing the primary class for the new characters. More precisely, the rotated characters are obtained by scanning each row of the starting bitmap and taking into account the extremes of each horizontal segment present in the row. The new extremes positions are calculated on the basis of the rotation applied. The vertical shift due to the rotation is neglected since the rotation angle is small. Therefore, only the new horizontal coordinates need to be estimated.

The network is thus forced to recognize (by definition) the transformed characters as belonging to the same class as the first ones. This technique significantly increased the recognition rate by more than one percent. This is probably due to the regularity of the output functions that interpolate between the added points. It may be noticed that rotation is applied to the training samples before preprocessing takes place. Therefore, the effective transformation consists of the rotation plus the preprocessing techniques applied.

3.2.2 The Testing Phase

It is possible to increase the network performance when an unknown pattern is presented to the neural network. The proposed technique, which has improved the performance of the MLP, consists of multi-presentation of the test set samples. This method is based on the implementation of an artificial test set according to the rotation applied to the training set.

For each character of the test set, the same preprocessing as used for the training set elements is employed. From each one, the two slightly rotated characters are determined: each of these representations of the same character is shown to the network, obtaining three output vectors.

The vector's components linked to the same output are added, thus obtaining the final output vector of the network. The character is classified according to the class associated with the largest component of the output vector. In practice, the two highest values, A_1 and A_2 in the output vector, are selected and the character is recognized as belonging to the class associated with the output A_1 if the following condition is satisfied:

$$A_1 - A_2 \geq R_{th} \tag{11}$$

where R_{th} is the user defined rejection threshold.

We used two different networks for uppercase recognition and for digit recognition. The rejection curve obtained for the uppercase and digit characters as a function of changing threshold is shown in Figure 6. The recognition and reject rates are normalized with respect to the whole test set size. The 1% misclassification rate is achieved with about a 9% rejection rate for uppercase recognition and about a 4% rejection rate for digit recognition. It should be noted that the comparison of the difference A_1-A_2 with a fixed rejection level (see the above relation) is particularly suitable for a digital implementation.

The answer of the neural network can be considered as the average of the three positions: the initial one, and those slightly rotated left and right. The arithmetic mean executes a low-pass filtering on the network's outputs, reducing the gradient of the function along the curve bound to the rotation applied close to the pattern shown. We would emphasize that the triple presentation of each pattern of the whole NIST test set increases the recognition rate by more than one percent.

4   Digital Implementation

The advantages offered by the proposed neural approach are not limited to better recognition and speed. The simplicity of the techniques employed leads to an architecture that is easily implemented on chips with classical VLSI techniques.

The digital architectural design aims to the best compromise between several constraints, such as high modularity of design, high density of neurons on chip, high recognition rate and speed. Taking these constraints into account:

- data input acts in a serial way: each time a three-bit datum flows into the chip;

- data processing acts in a parallel way among the neurons and in a serial way within each neuron;

- second layer processing is pipelined with first layer processing.

The proposed architecture is made up of two layers of neurons and is shown in Figure 7.

A first requirement is a compromise between the accuracy of the input data and the weight values, and the requirement to simplify the circuit that implements the dot products. Thus, we have chosen to reduce the matrix X into a 8*8 matrix whose elements are integers in the range of [-2, 2], so that for each one, three bits are sufficient.

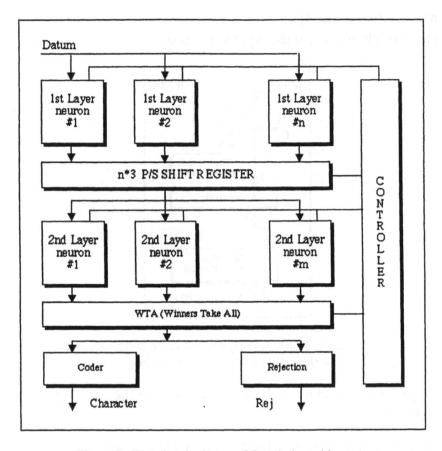

Figure 7. Functional scheme of the whole architecture.

This simple transformation leads to a new **N** matrix whose elements are obtained from the **M** ones (see Equation 10) by applying the following Equation:

$$n_{i,j} = \mathrm{int}\left(m_{i,j}/2\right) - 2 \qquad (12)$$

The simulations have shown that six bits for each weight of the first layer and five bits for each one of the second layer are sufficient for weights quantization. In this way the calculation of the product between each input value and the weight is very easy. In fact, the required operations are reduced to:

- a shift of the weight register if the input value is 2 and eventually a NOT operation if the input is -2;
- a NOT operation if the input value is -1;

- no operation if the input value is 1;
- clear weight register if the input value is 0.

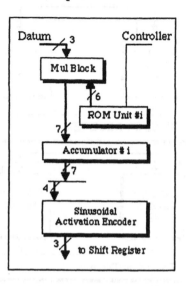

Figure 8. The block diagram shows the operations
carried out by a neuron of the first layer.

This operation is performed by the *Mul Block* in the first layer neuron of Figure 8.

A 7-bit accumulator is suitable for the required accuracy. In order to obtain a module of 128 pre-synaptic value and to code the first layer weights, six bits are sufficient to apply the following Equation:

$$W_{quantized} = round\left(W_{floating} \frac{128 * \Omega}{2\pi} \right) \qquad (13)$$

where $W_{quantized}$ is the new weight value, $W_{floating}$ the original weight, and Ω is the "frequency" of the sinusoidal function used in the training phase. (In our trials, $\Omega = 0.5$.) In order to quantify the second layer weights using 5 bits, we apply the following Equation:

$$W_{quantized} = round\left(W_{floating} \frac{15}{W_{max}} \right) \qquad (14)$$

where $W_{quantized}$ is the new value of the weights, $W_{floating}$ the original one and W_{max} the maximum value of the second layer weights.

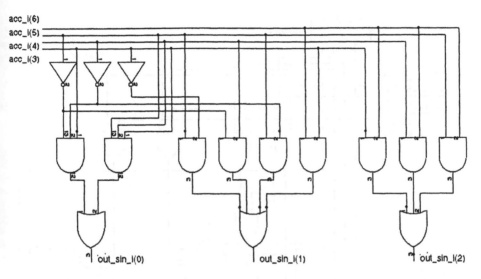

Figure 9. The sinusoidal activation encoder circuit of the ith neuron of the first layer. The accumulator's four MSB bits are codified obtaining second layer inputs.

A specific circuit (*Sinusoidal Activation Encoder*) processes the pre-synaptic activation value in order to obtain the related post-synaptic activation value. We use three bits to code the output. The allowed output values are again the -2, -1, 0, 1, 2, so the second layer will act in the same way as the first layer. The sinusoidal activation encoder is shown in Figure 9.

The second layer output is sent to the Winners Take All (WTA) circuit that individualizes, among a set of m numbers, each of which consists of p bits; the two greatest ones as the two greatest activation level units in a time independent by m and growing linearly with $2*p$.

If the classification is weak (i.e., the difference between the winning unit activation and the second classified unit activation is less than a prefixed quantity), a rejection circuit (see Figures 7 and 10) rejects the character to guarantee a better classification rate.

We have developed and tested digital implementation of the proposed neural architecture using the VHDL language [19] [21]. Each module, as well as the whole architecture, was built up and successfully tested.

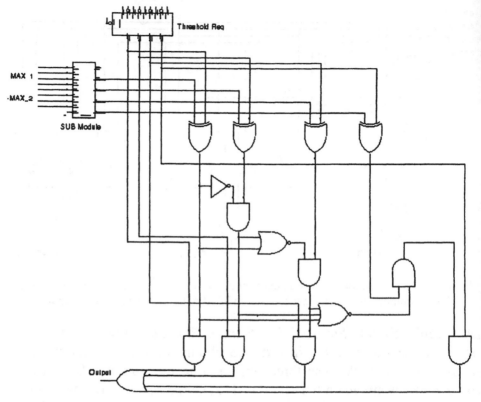

Figure 10. The rejection circuit acting on four bits registers. The Sub-Module makes the differences between the two greatest activation level units of the second layer. The Threshold Register contains an user prefixed threshold.

5   Experimental Results

The MLP networks used in the software experiments have the topology 64-256-26 for the uppercase character recognition task and 64-128-10 for the digits one. They are trained starting with randomly chosen initial weight values. The data used for training the network are those present in the NIST database. The training set consists of 44,951 elements for the uppercase character recognition task and 223,125 elements for the digit character recognition task.

In the first phase, only a few thousand characters are shown to the network in order to increase the learning process speed. After about ten epochs, the number of samples is increased until it includes the whole training set. By using this method the first subset fixes the weight values close to optimal, while the remaining samples are used for fine weight tuning. This technique has shown a great capability to avoid local minima [12].

MLP's performance was tested using 11,941 elements for the uppercase character recognition task and 58,536 elements for the digit character recognition task. The trials were conducted to emphasize the MLP's performance with varying activation functions for the hidden and output layers and to show the contribution given by each technique employed (enhancement of the training set, triple presentation of test samples).

The first experiments were about the optimal choice of the activation functions for the hidden and output layers. In this phase, heuristics H1 have been employed as preprocessing techniques.

Table 2 shows the software simulations recognition rate achieved by the MLP in the uppercase characters recognition task with the different activation functions of the hidden and the output layers.

These experimental results especially show the importance of the linear activation function in the output layer. In addition, the comparison between the learning speeds has also shown how the use of the sine function leads toward better performance.

Table 1-A. The proposed heuristics for the Preprocessing phase.

Topics	Advantages Pursued	Heuristic[1]
Pre-processing	A1. High speed	H1. This request is obtained by using the following simple heuristics:
	I. Rough independence from character thickness and size.	I. Application to the 128*128 starting binary matrix of the Equation (9) in order to obtain a normalized character into a 24*24 binary matrix.
	II. Reduction of character matrix dimension keeping the character features	II. Further reduction of the 24*24 binary matrix into a 8*8 integer matrix by application of Equation (10).

---

[1] Some adopted choice may require a theoretical explanation to justify their employment, however the results obtained encourage their use.

Table 1-B. The proposed heuristics for the Training phase.

Topics	Advantages Pursued	Heuristic[1]
Training	A2. Network invariance from slightly rotated input characters.	H2. Artificial enhancement of the training set by addition of new characters obtained from the starting ones by applying a rotation of an angle ±α.
	A3. Avoid the local minima of the network error function	H3. Use a sinusoidal activation function for the hidden neurons and a linear activation function for the output layer
	A4. Improve the location of the class separation surfaces and the learning speed.	H4. Make two learning steps: I. in the first step a linear function with a 0.5 slope is used; II. in the second step a linear function with a 0.1 slope is used.

Table 1-C. The proposed heuristics for the Testing phase.

Topics	Advantages Pursued	Heuristic[1]
Testing	A5. Improve the network performance in the recognition of unknown patterns.	H5. Show to the network an unknown pattern enhanced with two new characters obtained from the primary one by applying a rotation of an angle ±α

Table 2. MLP's performance for uppercase characters by varying the activation functions

MLP sigmoid-sigmoid	MLP sigmoid-linear	MLP sine-linear
91.6%	93.1%	93.4%

In order to emphasize the contributions of the different heuristics, a distinction has been made. Looking at Tables 1-A, 1-B and 1-C, we denote with

- MLP-1, the neural network implementing the heuristics H1, H3 and H4;
- MLP-2, the neural network implementing the heuristics H1, H2, H3 and H4;
- MLP-3, the neural network implementing each proposed heuristic.

The performance reported in Table 3 for uppercase character and digits recognition is very satisfactory. A neural network like the MLP-3 with an architecture 64-128-10 has been employed for digits recognition obtaining a 97.02% recognition rate.

Table 3. Uppercase character recognition percentages
achieved with the implemented MLP networks.

Neural Network	Uppercase recognition rate (%) (zero rejection rate)	Digit recognition rate (%) (zero rejection rate)
MLP-1	93.4	/
MLP-2	94.82	96.45
MLP-3	**96.2**	**97.02**

The first simulations of the digital architecture, realized entirely in VHDL, have shown that the recognition performance is close to the best one reported on Table 3. A recognition rate of about 95.95% is obtained for the handwritten digits recognition task using the same neural network topology (64-128-10).

Performance losses are mainly due to the rough quantization of the activation function. No improvement is gained using more than 6 bits in order to represent the weights of the neural network.

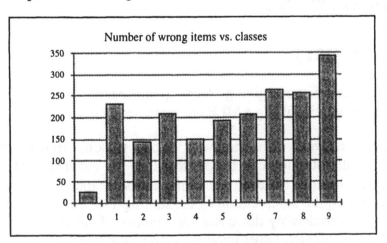

Figure 11. Error rate for each class
in the digits handwritten characters recognition task.

Figure 11 reports the absolute error we obtained for each class. Readers can see the more difficult classes recognized are the "9" and "7" classes, while the easiest class recognized is the "0" class.

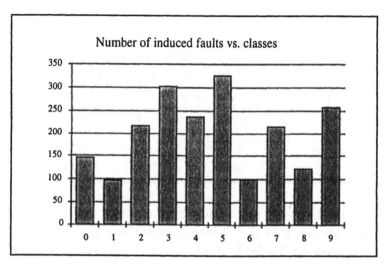

Figure 12. Error inducing capacity for each class
in the digits handwritten characters recognition task.

In Figure 12, we report a diagram showing the error that each class causes in other classes. The most confusing class for the handwritten digits recognition tasks are the "5" and "3" classes, while the least confusing is the "1" class.

6   Summary

Innovative techniques giving high performance for handwritten character recognition using a MLP are presented in this chapter.

Particularly, network invariance with respect to some transformations such as small rotations can be produced by an enhancement of the training set, leading to lower gradient values of the output function [11]. Moreover, the compression of the starting 128*128 matrix into an 8*8 matrix considerably decreases the number of input units and the network's complexity. Other choices, such as the triple presentation of the test samples and the use of sinusoid and linear activation functions have contributed to the high recognition rates.

The experimental results are very interesting in terms of both recognition rate and speed. We obtained 97.02% accuracy on the handwritten digit characters and 96.20% accuracy on the handwritten uppercase characters.

The employed techniques can be easily implemented in a fully digital architecture [17] [18] [19]. The first simulations of the derived digital architecture, using a VHDL simulator, have shown that the recognition rate decreases by about 1% on the handwritten digit characters.

References

[1] Kovàcs, M., Guerrieri, R., Baccarani, G. (1994), "A Novel Metric for Nearest-Neighbor Classification of Hand-Written Digits," *Proc. of 11th IAPR International Conference on Pattern Recognition*, pp. 96-100.

[2] Knerr, S., Personnaz, L., Dreyfus, G. (1992), "Handwritten Digit Recognition by Neural Networks with Single-Layer Training," *IEEE Trans. on Neural Networks*, **3**, 6, pp. 962-968.

[3] Zhang, M., Suen, C.Y., Bui, T.D. (1994), "An Optimal Pairing Scheme in Associative Memory Classifier and its Application in Character Recognition," *Proc. of 11th IAPR International Conference on Pattern Recognition*, pp. 50-53.

[4] Specht, D.F. (1990), "Probabilistic Neural Networks," *Neural Networks*, Vol. **3**,1, pp. 109-118.

[5] Rosenblatt, F. (1959), *"Principles of Neurodynamics,"* New York, Spartan Books.

[6] Minsky, M., Papert, S. (1969), *"Perceptrons,"* MIT Press, Cambridge, MA.

[7] Hornik, K., (1989), "Multilayer Feedforward Networks are Universal Approximators," *Neural Networks*, **2**, pp. 359-366.

[8] Powell, M.J.D. (1968), "Restart Procedures for the Conjugate Gradient Method," *Mathematical Programming*, **12**, pp. 241-254.

[9] Barnard, E., Casasent, D. (1991), "Invariance and Neural Nets," *IEEE Trans. on Neural Networks,* **2**, 5, pp. 498-508.

[10] Giles, C.L., Griffin, R.D., Maxwell, T. (1988), "Encoding Geometric Invariances in Higher Order Neural Networks," *Neural Information Processing Systems*, D.Z. Anderson, Ed., Denver, CO, pp. 301-309.

[11] Simard, P., Le Cun, Y., Denker, J., Victorri, B. (1994), "An Efficient Algorithm for Learning Invariances in Adaptive Classifiers," *Proc. of 11th IAPR International Conference on Pattern Recognition*, pp. 651-655.

[12] Gori, M., Tesi, A. (1992), "On the Problem of Local Minima in Backpropagation," *IEEE Trans. on PAMI*, **14**, 1, pp. 76-85.

[13] Stork, D.G., Allen, J.D. (1992.), "How to Solve the N-Bit Parity Problem with Two Hidden Units," *Neural Networks*, **5**, pp. 923-926.

[14] Gorse, D., Shepherd, A., Taylor, J.G. (1994), "A Classical Algorithm for Avoiding Local Minima," in *Proceedings of WCNN'94*, **III**, pp. 364-369.

[15] Lowe, D., Webb, A.R. (1991), "Optimized Feature Extraction and the Bayes Decision in Feed-Forward Classifier Networks," *IEEE Trans. on PAMI*, **13**, 4, pp. 355-364.

[16] Wilkinson, R.A., Geist, J., Janet, S., Grother, P.J., Burges, C.J.C., Creecy, R., Hammond, B., Hull, J.J., Larsen, N.J., Vogl, T.P., Wilson, C.L. (1992), "The First Optical Character Recognition Systems Conference," *Technical Report NISTIR 4912*, National Institute of Standards and Technology.

[17] Gioiello, G.A.M., Vassallo, G., Sorbello, F. (1992), "A New Fully Digital Neural Network Hardware Architecture for Binary Valued Pattern Recognition," *Proc. of the International Conference on Signal Processing Applications and Technology* (ICSPAT) Boston, pp. 705-708.

[18] Gioiello, G.A.M., Vassallo, G., Sorbello, F. (1993), "A New Fully Digital Feed-Forward Neural Network For Hand-Written Digits Recognition," *Proc. of VI Italian Workshop on Parallel Architectures And Neural Networks*, World Scientific, pp. 255-260.

[19] Perry, D.L. (1994), *VHDL*, McGraw-Hill Inc.

[20] Gioiello, G.A.M., Vassallo, G., Condemi, C., Sorbello, F. (1993), "A VLSI Digital Solution for Hand-Written Digit Recognition Using a Neural Network," *Proc. of DSP & CAES-93* Nicosia, Cyprus, pp. 141-145.

[21] Asheden, P.J. (1996), *The Designer's Guide to VHDL*, Morgan Kaufmann Publishers.

[22] Cybenko, G. (1988), *Approximations by Superpositions of a Sigmoidal Function*, Research Note, Computer Science Department, Tufts University.

Chapter 6:

Signature Verification Based on a Fuzzy Genetic Algorithm

SIGNATURE VERIFICATION BASED ON A FUZZY GENETIC ALGORITHM

J.N.K. Liu and **G.S.K. Fung**
Department of Computing
Hong Kong Polytechnic University
Hung Hom, Hong Kong
csnkliu@comp.polyu.edu.hk

The signature of an object can be regarded as some feature or characteristic specific to it, which shall be differentiated from that of other objects for verification purposes, including recognition, identification and authentication. These features or characteristics could be static (e.g., fingerprint, face profile) or dynamic, exhibiting some variations (e.g., speech, trembling property in signature handwriting) during transition change, which can be represented as a sequence of samples with periodic behaviour. The aim of this study is to automatically identify the important signature features and verify signatures with greater certainty. The use of fuzzy genetic algorithm (FGA) overcomes the traditional problems in feature classification and selection, providing fuzzy templates for the identification of the smallest subset of features. Results indicate that the presence of some features does create confusion and noise to the classifier for the verification process. The study suggests that FGA can be applied to select the key features and the class-dependent discretization method to discrete the selected features. The major advantages are that dynamic information of a signature can be hidden from other users and the comparison process is made very efficient through the signature feature discretization process.

0-8493-9807-X/99/$0.00+$.50

1   Introduction

A signature of an object is understood to have some features or characteristics specific to it, shall it be the traditional signature handwriting, the trailing smoke from the specific engine of some aircraft, the sonar signal of a moving object under the ocean, etc. The problem of signature verification lies in how we can specifically verify this object in terms of its unique features through different processes including recognition, identification and authentication. Given the complexity of possible variations in signature, how can we make use of the feature information, especially the dimensional variations (e.g., spatial, temporal), for solving the problem? In addition, the resource constraints do require the best use of crucial features that shall be properly selected from the domain of interest. This is the major step toward solving the optimization problems in the verification process [12].

In most of the existing signature verification methods, a feature set that has been regarded as crucial to signature verification was usually defined by the researchers' own experience. The appropriateness of the proposed feature set for some classification algorithms has rarely been studied and justified. As noted there are no solid procedures specified for identifying these important signature features. To tackle the feature selection problem, numerous approaches including probabilistic measures of class separability, stepwise forward and backward addition and/or elimination methods, branch and bound algorithm, and traditional genetic algorithm have since been proposed. However, they all suffer from some underlying assumptions such as the features independence, monotonicity condition, and sample complexity whose violation will lead to computational impracticality.

To overcome those problems in literature, we introduce an enhanced feature selection methodology using fuzzy genetic algorithm [7]. In general, we aim to produce a verification system with two main objectives:

(1) To reduce the possible features to a set of relevant uncorrelated key features in order to improve the correct classification rate.

(2) To provide means of finding a smaller set of features to reduce the time complexity of the algorithm as the effectiveness of most algorithms will depend on the dimensionality of the features space.

Thus, the reduction of the feature set means that smaller sample sets can be used with all the ensuring benefits in terms of data collection and processing time. This has direct influence to the successful completion of signature verification in many application areas.

2   Background

As noted in literature, most of the signature verification systems concentrated on handwritten signature area including the off-line and on-line processing of a signature [3], [17]. Typically, Leclerc and Plamondon [18] has provided an overview of achievements in both static and dynamic methods in order to solve problems in the design and implementation of signature verification systems. Dynamic programming techniques were introduced to determine a feasible solution for signature matching [27], [30]. Herbst and Liu [10] suggested the regional correlation algorithm to solve the problem of speed variation in signature writing. In general, these methods suffer from the requirement of sizable memory and large computational effort. There might be other research in artificial intelligence areas [11], [19], [20], [22] which addressed pattern matching problems using techniques that can be relevant to the verification process. Without any loss of generality, we shall concentrate on some domain which does exhibit static features and/or dynamic ones representing some behavioral properties for investigation. For example, the static features are those which shall be sensitive to the shape variation of signature (e.g., stroke density) while those of the dynamic ones shall be sensitive to the temporal variation of some properties during signature formation. This could be related to the linking sequence of signature segments, some writing behaviour inherited by an individual (e.g., trembling property), etc. Among these many problems, feature selection causes the most important concern in signature verification and is the focus of this study.

2.1   Problems Associated with Signature Verification

Many signatures are subject to two types of signature variability:

(1) Intraclass variability: the variation observed within a class of genuine signature specimens of an object (individual).
(2) Interclass variability: the variation which exists among different genuine signature classes.

In theory, intraclass scatter must be as low as possible and interclass large enough to be used for class separation. In practice classes are not obviously well separated. Therefore, signature verification cannot be considered as a trivial pattern recognition problem. It relies on selected features which are a set of transformations of the signature data that are able to differentiate genuine signatures among many possible choices.

Considering the complexity of feature selection, assume that Q is the original set of features with cardinality n, d is the desired number of features in the selected subset P such that $P \subseteq Q$, $J(P)$ is the feature selection criterion function for the set P. Without any loss of generality, a higher value of J is considered as an indication of a better feature subset. Jain and Zongker [12] described the problem of feature selection as finding a subset $P \subseteq Q$ such that $|P| = d$ and $J(P) = \max_{R \subseteq Q, |R| = d} J(R)$. An exhaustive approach to this problem would require the examination of all $\binom{n}{d}$ possible subsets of the feature set Q.

The number of possibilities grows exponentially, making an exhaustive search impractical for even moderate values of n. It was shown that no nonexhaustive sequential feature selection procedure could be guaranteed to produce the optimal subset. Also, it was shown that any ordering of the error probabilities of each of the 2^n feature subsets is possible [4].

In addition, the performance of a verification system is generally evaluated according to the error representation of a two-class pattern recognition problem, that is, with the type I (FRR: false rejection) and type II (FAR: false acceptance) error rates. These rates vary with the acceptance/rejection threshold. The ideal situation is summarized in Figure 1(a). It would occur if the "best features" were selected to separate completely true signatures from forgeries where applicable. In this case, no imitation or degeneracy would occur. In practical situations, as forgeries are not necessarily available, and as the feature selection problem has not been satisfactorily solved, the feature choice is not optimal and one often has to cope with curves as shown in Figure 1(b), where the four types of specimen occur. This figure also points out the difficulty of evaluating performances between different signature verification systems through type I and type II errors rates, since thresholds are different from one experiment to another and field tests are applied to different signature data sets.

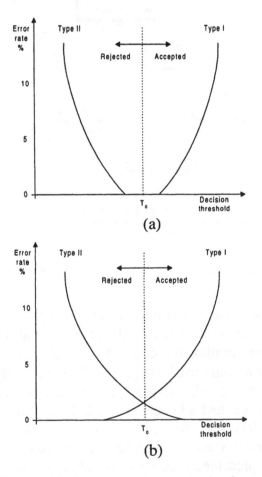

Figure 1. Class partitioning and error rate evaluation from similarity measure
(a) ideal case; (b) practical case.

Consequently, we need to derive a means of learning these types of signature variability, better formulation for the feature selection criteria, and an efficient searching algorithm in support of signature verification.

2.2   Feature Selection Algorithms

The taxonomy of available feature selection algorithms is presented in Figure 2. The methods are mainly divided into those based on statistical pattern recognition (SPR) techniques, artificial neural networks (ANN), and those using hybrid-type systems (HS).

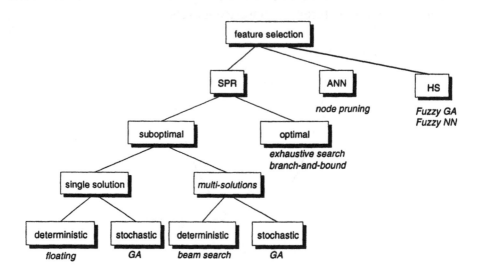

Figure 2. Taxonomy of feature selection algorithms.

As shown in the figure, the first group of methods begins with a single solution (i.e., a feature subset) and features are iteratively added or removed until some termination criterion is met. These are also referred to as the commonly used "sequential" methods for performing feature selection. They can be divided into two categories, those that start with the empty set and add features and those that start with the full set and delete features. Since they do not examine all possible subsets, these algorithms are not guaranteed to produce the optimal result. Kittler [14] presented a comparative study of these algorithms and the optimal branch-and-bound algorithm using a synthetic two-class Gaussian data set. Pudil et al. [25] updated this study by introducing them to "floating" selection methods.

Siedlecki and Sklansky [28] applied a best-first search and beam search, respectively, in the space of feature subsets. Both these methods maintain a queue of possible solutions. They treat the space of subsets as a graph, called a "feature selection lattice", (where each node represents a subset and an edge represents the containment relationship) and then apply any one of a number of standard graph-searching algorithms.

Narendra and Fukunaga [24] proposed a branch-and-bound feature selection algorithm which can be used to much more quickly find the optimal subset of features than the exhaustive search. However, Jain and Zongker [12] noted that the branch-and-bound procedure requires the feature selection criterion function to be monotonic. It implies that the addition of new features to a feature subset can never decrease the value of the criterion function. This is not true in some small sample size situations. In addition, the branch-and-bound method is still impractical for problems with very large feature sets, due to combinatorial explosion of complexity in the search space.

Hamamoto et al. [9] demonstrated that the branch-and-bound procedure performs well, even in cases where the feature selection criterion is nonmonotonic. A modification of the branch-and-bound algorithm [24], called BAB^+, was found to be outperforming the original algorithm [30]. This modification basically recognizes those subtrees that consist of a single path from the root to a terminal node, and save intermediate evaluations by immediately skipping the search forward to the appropriate terminal node.

Liu and Lee [21] used a multilayer feedforward network with a backpropagation learning algorithm for pattern classification. They defined a "node saliency" measure and presented an algorithm for pruning the least salient nodes. The pruning of input nodes is equivalent to removing the corresponding features from the feature set. It reduces the complexity of the network after it is trained, and at the same time, develop both the optimal feature set and the optimal classifier. The saliency of a node is the sum of the increase in error represented by a squared-error cost function, taking into account all the training patterns, as a result of removing that node. The node saliency is approximated with a second-order expansion and then computed the value by finding the appropriate derivatives in a backpropagation fashion. This method is more practical in computing the saliency, as it requires only one pass through the training data (versus one pass per node).

For feature selection, the node-pruning-based methodology first trains a network, and then removes the least salient node (input or hidden). The reduced network is trained again, followed by the removal of the least salient node. This procedure is repeated until the desired tradeoff between the classification error and the size of the network is achieved.

In late 1980s, the genetic algorithm (GA) was used to optimize feature sets for classification systems. In the application of GA to feature selection, each chromosome represents a subset of features, that is, the presence or absence of kth feature is represented by the kth bit of the chromosome. The fitness function of the individuals is defined by the error rate and number of selected key features. After a certain predefined number of generations, a population of sub-optimal solutions resulted. An advantage of this search method is that it tends to produce "good" solutions quickly rather than trying to find the "optimal" solution.

A recent development for feature selection considers the integration of fuzzy classification and genetic algorithm. This is to provide fuzzy templates for the identification of the smallest subset of features [6], [7]. In addition, hybrid systems including neural networks, active contour model and genetic algorithm have been used to investigate the invariant features of facial and handwritten images [20], [21]. The integration provides an improvement in feature classification and discrimination of closely resembled images.

3   The Fuzzy Genetic Algorithm

The traditional GA was originally designed only for searching through the discrete space. To overcome this limitation, some coding procedures are needed to represent a continuous search space. However, the resolution of the search space, that we may not know in advance, has to be defined by the coding scheme. Therefore, a more direct and efficient approach is to fuzzify the genetic algorithm. The resulting Fuzzy Genetic Algorithm (FGA) tends to be more efficient and more suitable for some applications [15].

Basically, there are two ways to fuzzify the GA. One way is to fuzzify the gene pool and the associated coding of chromosomes. The other is to fuzzify the operations on chromosomes. These two ways may be applied at the same time. In GA, a chromosome is represented by a finite sequence of 0s and 1s. In FGA, the gene pool is fuzzified by representing the chromosome with a finite sequence of real numbers ranging from 0 to 1. Buckley and Hayashi [2] fuzzified the GA by just extending the gene pool from {0, 1} to [0, 1]. However, a more genuine fuzzification process is needed so that the operations on the chromosomes can also be fuzzified. Sanchez [26] proposed a soft crossover technique for FGA, in which fuzzy template is used in addition to crossover point. This enables progressive exchange at the crossover point. Based on the types of operations, we can classify FGA into two categories, hard and soft. In hard FGA, the new offspring is created by exchanging the segments of the parents at the crossover point. This is identical to the GA procedure. In soft FGA, fuzzy template is defined to enable gradual crossover of two parents at the crossover point. Klir and Yuan [15] claimed that this kind of FGA is more efficient, robust and better attuned to some applications than its classical, crisp counterparts.

The FGA generalized for both hard and soft categories is defined in details as follows:

---

(1) **Population initialization.**

$\Sigma = \{\overline{X_1}, ..., \overline{X_n}\}$, an initial population of size n is randomly generated from $[0, 1]^{N+1}$.

where $\overline{X_1} = (x_{10}, \cdots, x_{1N})$,

$\vdots$

$\overline{X_n} = (x_{n0}, \cdots, x_{nN})$, where x_{ij} is in $[0, 1]$, $1 \le i \le n$, $1 \le j \le N$.

(2) **Fitness calculation.**

$\theta_i = f(\overline{X_i})$, fitness is computed for each $\overline{X_i}$, $1 \le i \le n$.

$T = \theta_1 + ... + \theta_n$, the total sum,

$s_k = \theta_1 + \ldots + \theta_k$, the partial sums, $1 \leq k \leq n$,

$I_1 = [0, s_1)$, $I_i = [s_{i-1}, s_i)$, $2 \leq i \leq n\text{-}1$, $I_n = [s_{n-1}, s_n]$, intervals constructed from the partial sums.

(3)   **Parents population formation.**

w_i = a random number in $[0, T]$, $1 \leq i \leq n$,

If w_i is in I_j, $\overline{X_j}$ will be included in the parents population.

$\Psi = \{\overline{P_1}, \ldots, \overline{P_n}\}$, the parent population,

Note: Duplications are allowed.

(4)   **Crossover.**

Two children are produced from each pair of parents in parent population via the 'crossover' operator,

p = the probability of a crossover in $[0, 1]$,

The simple fuzzy crossover process of the parents, $\overline{P_1}$ and $\overline{P_2}$, is described as follows:

x = a random number in $[0, 1]$,

If $x \leq p$, then perform crossover on $\overline{P_1}$ and $\overline{P_2}$,

r = a random integer indicating the crossover position $[0, N\text{-}1]$,

$\overline{F} = (f_1, \ldots f_N | f_1 = 1, f_N = 0, i < j \Rightarrow f_i \geq f_j, f_r \approx 0.5)$, *fuzzy template*,

$\overline{\widetilde{F}} = (\tilde{f}_1, \ldots \tilde{f}_N | \tilde{f}_i = 1 - f_i)$,

the two children become

$\overline{P_1}' = (\overline{P_1} \wedge \overline{F}) \vee (\overline{P_2} \wedge \overline{\widetilde{F}})$,   where $\wedge$ and $\vee$ are min and max operations.

$\overline{P_2}' = (\overline{P_1} \wedge \overline{\widetilde{F}}) \vee (\overline{P_2} \wedge \overline{F})$,

If $x > p$, then there is no change in $\overline{P_1}$ and $\overline{P_2}$,

i.e., the two children, $\overline{P_1}'$ and $\overline{P_2}'$, are identical to their parents $\overline{P_1}$ and $\overline{P_2}$.

Note: There are 2 types of crossover operation: *Hard* and *Soft*,

hard fuzzy template: sharp gene exchange, e.g., $\overline{F} = \langle 1, \ldots, 1, 0, \ldots 0 \rangle$,

soft fuzzy template: progressive gene exchange, e.g.,

$\overline{F} = \langle 1, \ldots, 1, 0.8, 0.5, 0.2, 0, \ldots 0 \rangle$.

(5)   **Mutation.**

The mutation operation is performed position by position on each child $\overline{P_i}'$,

q = the probability of a mutation in $[0, 1]$,

w_i = a random number in $[0, 1]$, $0 \leq i \leq N$,

Consider $\overline{P_1}'$, if $w_i \leq q$, then mutate the ith position in $\overline{P_1}'$,

if $w_i > q$, then no change in the ith position, $0 \leq i \leq N$.

For example, $w_2 < q$ and $w_i > q$, $0 \leq i \leq N$, $i \neq 2$,

p_{12}^{*}, a random number in $[0, 1]$,

$\overline{P_1}'$ becomes $\overline{P_1}' = (p_{10}, p_{11}, p_{12}^{*}, ..., p_{1N})$ with p_{12} replaced by p_{12}^{*}.

(6)   Loop.
The loop, steps 2 to 5, is done L times, where L is the maximum number of iterations.

---

4   Feature Selection Problems

Recently, the FGA [6], [7] was introduced for feature selection problems (FSP). The selection of features associated with each individual is based on the fitness-proportionate selection in which parents are chosen from the population. Results demonstrate that the operation using soft crossover will improve the searching power through the multidimensional feature space. It is noted that there are weaknesses or limitations with the GA approach to FSP, but these limitations can be resolved by the FGA. First, there are only 2 different selection degrees or categories for features in the GA: either being completely selected or completely rejected. It is obvious that it would be desirable to rank the degree of importance of the features. For the important features, the degree of importance should be close to 1; on the other hand, for the irrelevant features, the degree of importance should be close to 0; for the in-between cases, the degree of importance should be somewhere between 0 and 1 depending on their importance. This FGA approach can further improve the performance of the system because it is able to search through the continuous space by new chromosome type and operations. Second, over a range of genetic algorithm settings, i.e., the probability of crossover and that of mutation of the GA, the performance of the system varies significantly. This may be caused by the incomplete search of the feature space by the restricted information available. We believe that with the extra information of the degree of importance of the features in the FGA, the search space could be searched more efficiently and robustly.

Basically, there are two versions of the FSP which address the specific objective and lead to a distinct type of optimization.

(1) The objective is to find a subset that gives the lowest error rate of a classifier and the problem leads to unconstrainted combinatorial optimization with the error rate as the search criterion.

(2) The objective is to find the smallest subset of features for which the error rate is below a given threshold, and the problem leads to a constrainted combinatorial optimization task in which the error rate serves as a constraint and the number of features is the primary search criterion.

The present study focuses on the second type of optimization. The smallest subset of features is sought such that the classifier's performance would not deteriorate below a certain specified level. The fitness function is defined as follows:

$$a_i = \{\alpha_{i1}, \ldots, \alpha_{id}\}$$

$$pen(e) = \frac{\exp((e - thres) / m) - 1}{\exp(1) - 1}.$$

$$l(a_i) = \sum_{j=1}^{d} \alpha_{ij} | \alpha_{ij} \geq 0.5,$$

$$J(a_i) = l(a_i) + pen(err(a_i)),$$

$$\prod = \{a_1, \ldots, a_n\},$$

$$f(a_i) = (1 + \varepsilon) \max_{a_j \in \prod} J(a_j) - J(a_i),$$

where a_i is the feature selection vector,

α_{ij} is the degree of importance of ith feature of the jth feature selection vector,

$0 \leq \alpha_{ij} \leq 1$,

= 0.0 if it is completely excluded from the subset,

= 1.0 if it is completely included in the subset,

d is the number of original features,

pen denotes the partial penalty function,

e denotes the error rate,

$thres$ denotes the feasibility threshold,

m denotes the scale factor (i.e., tolerance margin),

J denotes the total penalty function,

err denotes the error rate function,

l denotes the summation of the degree of included features which has a degree of importance ≥ 0.5 in a_l,

f denotes the fitness function,

ε is a small positive constant,

$\prod$ denotes a population $\{a_1, \ldots, a_n\}$ of feature selection vectors.

The feature selection vector, a_i, is defined by the degree of importance, α_{ij}, of each vector. In GA, α_{ij} is an integer, either 0 or 1, that is, accept or reject. In FGA, α_{ij} becomes a real number, from 0 to 1, i.e., from the very irrelevant to the very important. l is the summation of the degree of included features, which is larger than 0.5. All the features with their degrees of importance smaller than 0.5 are excluded. The error rate is calculated by distance-weighted Fifth Nearest Neighbor (5-NN) classification method using Euclidean distance matrices [13]. The classifier was trained by the "leave-one-out" method since limited data samples were available.

There are four properties of the penalty function that make it suitable for feature selection purpose. First of all, a small reward is given to the feature subsets for which the error rate is under the feasibility threshold. Also, at the same level, the adaptability of the feature subsets is judged by the error rates associated with them. Moreover, a small penalty, in the range 0 to 1, is given to the feature subsets when the error rates fall between *thres* and *thres+m*. Finally, a high penalty, *pen*, (over 1) is given to the feature subsets with the error rate over *thres+m*, so these subsets could not compete with the subsets which have an extra feature.

5   Experiments with Signature Data

In order to demonstrate our new FGA approach in FSP involving signature data, three sets of features were tested. The first and second sets are handwritten signature features originally proposed in [6], of which one contains 91 features, while the other 69 normalized features extended and amended from the set proposed in [16]. The third set contains 127 facial features extended from [1].

5.1   Original Handwritten Signature Set

Signature data with spatial and temporal properties were collected with a CalComp Drawing Board III and a Pressure Cordless Pen. The sampling rate was set to 100 samples per second. There were 31 volunteers involved in the sampling of 310 signatures, that is, 10 signature samples were collected from each volunteer to minimize the biasing effect. Eight signals in the function of time were collected or computed for each signature sample: the pen tip position along the x and y axes of the tablet, (x, y); the pen tip distance above the drawing surface of the tablet, (z); the tilt angle of the pen against the x and y axes of the tablet, (tx, ty); the pressure exerted on the pen tip, (p); and the pen tip velocity along the x and y axes of the tablet, (vx, vy). A sample of the time-dependent signature sample is given in Figure 3.

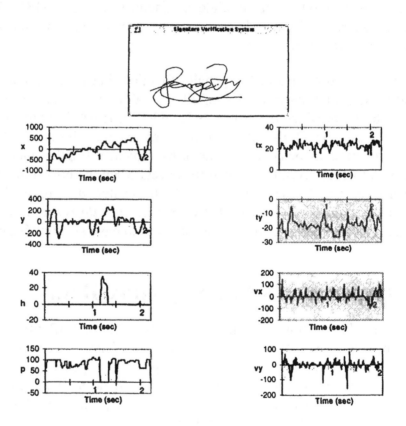

Figure 3. A sample of handwritten signature.

Based on the 44 original signature features set proposed in [5], we have expanded and modified it to the set with 91 features. They can all be extracted from the eight different signals. These 91 features are defined as in Table 1. To solve the position variance of the signature samples, the x and y are shifted by the mean of x and y, respectively. Also, since all the values of z and p signals are always positive, the features that consider positive and negative signals separately will be omitted in the original feature set. However, because we found that the ranges of the values of the 91 features can vary a lot, this will bias the classifier to the features with larger ranges. Therefore, to avoid this scaling problem, all the values of the features were scaled to standard unit deviation.

Table 1. The 91 original features of handwritten signature

FEATURE NUMBER								FEATURE
X	Y	Z	TX	TY	P	VX	VY	
-	-	21	29	42	55	63	76	mean
1	11	22	30	43	56	64	77	standard deviation
2	12	23	31	44	57	65	78	minimum
3	13	24	32	45	58	66	79	maximum
4	14	-	33	46	-	67	80	average absolute
5	15	-	34	47	-	68	81	average positive
6	16	-	35	48	-	69	82	number of positive samples
7	17	-	36	49	-	70	83	average negative
8	18	-	37	50	-	71	84	number of negative samples
9	19	25	38	51	59	72	85	number of zero-crossing
-	-	26	39	52	60	73	86	maximum - scaled mean
10	20	27	40	53	61	74	87	maximum - minimum
-	-	28	41	54	62	75	88	scaled mean - minimum
89								total time
90								time up / total time
91								time down / total time

5.2   Normalized Handwritten Signature Set

A set of 69 normalized candidate features was extended and amended from the 49 normalized feature set in [16]. As claimed in [16], these 69 normalized features should be less sensitive to intra-personal variation in size, velocity, pressure, and pen tilt angles of genuine signatures. The 69 normalized features are described in the Appendix to this chapter. For instance, one of the normalized handwritten signature features is the ratio of total pen down duration to total signing duration.

5.3   Facial Signature Set

A facial signature is a set of facial features extracted from 400 different grayscale images of 40 different subjects (10 different images for each subject). These facial images were obtained from the ORL Database, the Olivetti Research Laboratory, Cambridge, U.K. They were originally used in a face recognition project. These facial images were captured under a certain degree of variations in lighting, subject position and orientation, containing far more information for recognizing faces. In order to reduce the domain of the raw input data for higher-level processes in recognition, important facial features need to be selected so that the most essential ones are used.

Three basic types of features were used in the experiments. They were the normalized size of facial features, the relative position of the feature to a fixation point, and the relative intensity. These features are the extended set of those proposed in [1]. Besides the normalized feature sizes used in the paper, which indicated the normalized sizes of different important facial components such as eyes and mouth, global relative positions of features to a fixation point and relative intensity differences were added. Fixation point [29] is the point that statistically a person looks at when a face is presented. This point can be used as a reference point to other important features. Usually the nose is the first point that people will look at, and is thus the fixation point. The relative positions of different feature points to the nose were used, and then normalized to avoid position variation of the subject.

Although people are able to recognize facial signatures with a large range of intensity variation, we believe that the relative intensities among different features can be used as additional information. The intensities of eyebrows, eyes, lips, cheeks and hair relative to that of the nose were used. Figure 4 shows the feature points selected for determining these facial features, while Figure 5 shows one of the facial images used and its corresponding feature points manually selected.

In total we have 127 different defined features to be selected as shown in Table 2. 24 of them are normalized feature lengths/sizes of local features, 94 are feature positions relative to the fixation point, and 9 are relative intensities.

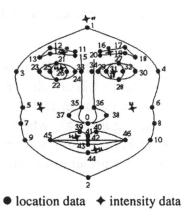

● location data ✦ intensity data

Figure 4. Facial points.

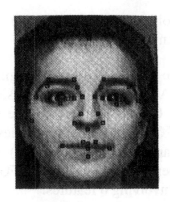

□ location data

Figure 5. Sample facial image.

Table 2. Features description of facial signature

Features Type	Number of Features	Facial Points Used
Normalized size of features in facial signature	24	Calculation of the facial components size from facial points # 0~46, e.g., width of mouth.
Relative position to Fixation Point	94	x- and y-values of points # 1~46 relative to point # 0.
Relative Intensity	9	Intensity points # 47~55.
Total	**127**	

6 Simulation Results and Discussions

Two feature selection schemes, the traditional genetic algorithm (TGA) approach and FGA approach, were investigated in this research. In the TGA approach, chromosomes were represented by the binary code and the classification was processed by the (5,3)-Nearest Neighbor method. In the FGA approach, chromosomes were extended and represented by sequences of real numbers ranging between 0 and 1, soft fuzzy template was used in the crossover operation, and the classification was processed by the Fuzzy 5-Nearest Neighbor method. For each scheme, experiments were performed over a range of settings, including the probability of crossover, the probability of mutation, feasibility threshold and tolerance margin. These experiments were executed for 500 generations.

For the database using 91 original signature features, the performance of TGA and FGA with the feasibility threshold set to 2% are shown in Figure 6. The best cases of the total penalty, J, of each generation for both TGA and FGA approaches are plotted in Figure 6(a), while the worst are plotted in Figure 6(b). For example, in Figure 6(a), the lowest total penalty achieved by the FGA is equal to 7.2, with an error rate of 4.2% involving 6 features.

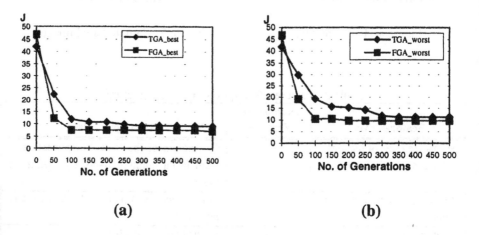

(a)   (b)

Figure 6. The performance comparison of TGA and FGA for 91 original signature features by (a) the best cases (b) the worst cases of the total penalty, J, after certain number of generations are processed.

For the database using 69 normalized signature features, the performance of TGA and FGA with the feasibility threshold set to 2% are shown in Figure 7. The best cases of the total penalty, J, of each generation for both TGA and FGA approaches are plotted in Figure 7(a), while the worst are plotted in Figure 7(b). For example, in Figure 7(a), the lowest total penalty achieved by the FGA is equal to 12.3, with an error rate of 5.2% involving 10 features.

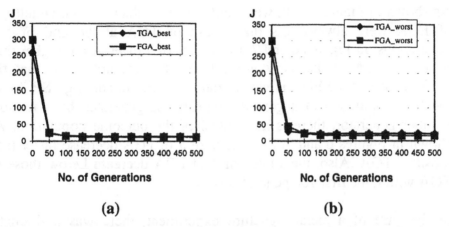

(a) **(b)**

Figure 7. The performance comparison of TGA and FGA for 69 normalized signature features by (a) the best cases (b) the worst cases of the total penalty, J, after certain number of generations are processed.

For the database using facial features, the performance of GA and FGA with the feasibility threshold set to 15% are shown in Figure 8. The best cases of the total penalty, J, of each generation for both GA and FGA approaches are plotted in Figure 8(a), while the worst are plotted in Figure 8(b). For example, in Figure 8(a), the lowest total penalty achieved by the FGA is equal to 29.21, with an error rate of 17.25% involving 28 features.

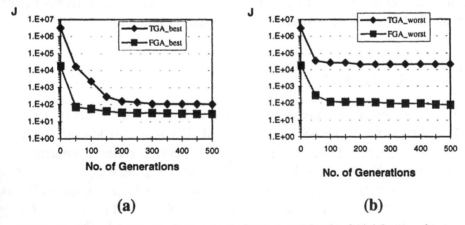

(a) **(b)**

Figure 8. The performance comparison of GA and FGA for facial features by (a) the best cases (b) the worst cases of the total penalty, J, after certain number of generations are processed.

As shown in Figures 6 - 8, we found that the values of total penalty, J, of FGA were always less than those of TGA at the end of 500 generations. In Figures 6 and 7, the values of J of FGA are higher than those of TGA at the early generations. It does not mean that the performance of the FGA is worse than the TGA in solving FSP. Since the first population of a genetic algorithm is generated by a random number generator, there is no control of the outcome population. As shown in Figures 6 and 7, the values of J of FGA decrease faster than those of TGA. Also, the values of J of FGA decrease below those of TGA within the first 100 generations.

In the case of a facial signature experiment, there was a dramatic improvement of FGA over TGA. However, the experiment using 69 normalized handwritten signature features gave a very little improvement. Therefore, the performance of the FGA might have been affected by the experimental database used. For all three testing feature sets, FGA has better solution than TGA in solving FSP at the end of 500 generations.

The performance of a genetic algorithm is judged by how fast it searches through the space, i.e., how fast it improves the solutions and how good the solution is at the end of all generations. FGA is able to improve the solutions faster than TGA and FGA has better solution than TGA at the end of all generations. Therefore, FGA performs better than TGA in our three experimental signature sets.

7   Summary

In this research, we presented a methodology called fuzzy genetic algorithm to solve the FSP in support of signature verification. Feature Selection is one of the methods used to reduce today's enormously complex problems. This has recently become a very active research area; however, we found that the current traditional genetic algorithm is insufficient to solve the feature selection problem. In this study, a fuzzy genetic algorithm was formally defined and proposed. There are two major advantages of fuzzy genetic algorithm over traditional genetic algorithm. First, the degree of importance for each feature within the key signature set could be evaluated. Second, the performance of the classifier could be improved. Three data sets including 91 handwritten features, 69 normalized handwritten features, and 127 facial features were tested by the proposed Fuzzy Genetic Algorithm to select the key feature subsets. Experimental results have confirmed the power and robustness of the Fuzzy Genetic Algorithm. It can facilitate the retrieval and identification of key information. The approach will keep dynamic information of a signature implicit in the code which is generated from some class-dependent discretization of the selected features [8]. Currently, we are studying the generic integration of information theory with Fuzzy Genetic Algorithm in applying to Feature Selection Problem. The solution to some domain specific problems has yet to be investigated and generalized for different signature verifications.

Acknowledgments

The facial images used for this study were provided courtesy of Olivetti Research Laboratory, U.K. We are also grateful to Dr. Rynson Lau and his team for their comments and contributions to the preparation of experimental data.

Appendix

Definitions (to ease symbol representation, $a \equiv tx$ and $b \equiv ty$.)

T_s	= Total signing duration	Δ_z	= total shift of z in pen down
T_w	= Total pen down duration	Δ_a	= total shift of a in pen down
$t()$	= duration in total time	Δ_b	= total shift of b in pen down
$t_u()$	= duration in pen up time	x_o	= $x(1^{st}$ pen down)
u	= number of pen ups	y_o	= $y(1^{st}$ pen down)
A_{min}	= $(x_{max}-x_{min}) \times (y_{max}-y_{min})$ in pen down	z_o	= $z(1^{st}$ pen down)
L	= signature length	a_o	= $a(1^{st}$ pen down)
$\overline{v}$	= average writing speed	b_o	= $b(1^{st}$ pen down)
$\overline{v_x}$	= average x component of writing speed	x_{end}	= x(last pen up)
$\overline{v_y}$	= average y component of writing speed	y_{end}	= y(last pen up)
$\overline{v_z}$	= average z component of writing speed	z_{end}	= z(last pen up)
v_{max}	= maximum writing speed	a_{end}	= a(last pen up)
v_x	= x component of writing speed	b_{end}	= b(last pen up)
v_y	= y component of writing speed	x_{max}	= maximum x
v_z	= z component of writing speed	y_{max}	= maximum y
v_{xmax}	= maximum x component of writing speed	z_{max}	= maximum z
v_{ymax}	= maximum y component of writing speed	a_{max}	= maximum a
v_{zmax}	= maximum z component of writing speed	b_{max}	= maximum b
v_{xmax}	= minimum x component of writing speed	x_{min}	= minimum x
v_{ymax}	= minimum y component of writing speed	y_{max}	= maximum y
v_{zmax}	= minimum z component of writing speed	z_{min}	= minimum z
Δ_x	= total shift of x in pen down	a_{min}	= minimum a
Δ_y	= total shift of y in pen down	b_{min}	= minimum b

69 Normalized Feature Set

1. T_w/T_s
2. u
3. $t(v_{max})/T_w$
4. $\overline{v}/v_{max}$
5. $t(+ve\ slope)/t(-ve\ slope)$
6. $[(x_{max}-x_{min})/(y_{max}-y_{min})]/[\Delta_x/\Delta_y]$
7. $std.\ dev.(x/\Delta_x)$
8. $std.\ dev.(y/\Delta_y)$
9. $A_{min}/(\Delta_x \times \Delta_y)$
10. L/A_{min}
11. $t(v_x=0)/T_s$
12. $t(v_x>0)/T_s$
13. $t(v_x<0)/T_s$
14. $t_u(v_x>0)/T_s$
15. $t_u(v_x<0)/T_s$
16. $t(v_{xmax})/T_s$
17. $t(v_{xmin})/T_s$
18. $\overline{v_x}/v_{xmax}$
19. $v_{xmin}/\overline{v_x}$
20. $(x_o-x_{max})/\Delta_x$
21. $(x_o-x_{min})/\Delta_x$
22. $(x_{end}-x_{max})/\Delta_x$
23. $(x_{end}-x_{min})/\Delta_x$
24. $t(v_y=0)/T_s$
25. $t(v_y>0)/T_s$
26. $t(v_y<0)/T_s$
27. $t_u(v_y>0)/T_s$
28. $t_u(v_y<0)/T_s$
29. $t(v_{ymax})/T_s$
30. $t(v_{ymin})/T_s$
31. $\overline{v_y}/v_{ymax}$
32. $v_{ymin}/\overline{v_y}$
33. $(y_o-y_{max})/\Delta_y$
34. $(y_o-y_{min})/\Delta_y$
35. $(y_{end}-y_{max})/\Delta_y$
36. $(y_{end}-y_{min})/\Delta_y$
37. $t(v_z=0)/T_s$
38. $t(v_z>0)/T_s$
39. $t(v_z<0)/T_s$
40. $t_u(v_z>0)/T_s$
41. $t_u(v_z<0)/T_s$
42. $t(v_{zmax})/T_s$

43. $t(v_{zmin}) / T_s$

44. $\overline{v_z} / v_{zmax}$

45. $v_{zmin} / \overline{v_z}$

46. $t(v_p=0) / T_s$

47. $t(v_p>0) / T_s$

48. $t(v_p<0) / T_s$

49. $t_u(v_p>0) / T_s$

50. $t_u(v_p<0) / T_s$

51. $t(v_{pmax}) / T_s$

52. $t(v_{pmin}) / T_s$

53. $\overline{v_p} / v_{pmax}$

54. $v_{pmin} / \overline{v_p}$

55. $(p_o-p_{max})/\Delta_p$

56. $(p_o-p_{min})/\Delta_p$

57. $(p_{end}-p_{max})/\Delta_p$

58. $(p_{end}-p_{min})/\Delta_p$

59. $(a_o-a_{max})/\Delta_a$

60. $(a_o-a_{min})/\Delta_a$

61. $(a_{end}-a_{max})/\Delta_a$

62. $(a_{end}-a_{min})/\Delta_a$

63. $(b_o-b_{max})/\Delta_b$

64. $(b_o-b_{min})/\Delta_b$

65. $(b_{end}-b_{max})/\Delta_b$

66. $(b_{end}-b_{min})/\Delta_b$

67. $[(a_{max}-a_{min})/(b_{max}-b_{min})]/[\Delta_a/\Delta_b]$

68. $std.\ dev.(a/\Delta_a)$

69. $std.\ dev.(b/\Delta_b)$

References and Further Reading

[1] Brunelli, R. and Poggio, T. (1992), "Face Recognition through Geometrical Features," *Proc. ECCV '92*, pp. 792-800.

[2] Buckley, J.J. and Hayashi, Y. (1994), "Fuzzy Genetic Algorithm and Applications," *Fuzzy Sets and Systems*, Vol. 61, pp. 129-136.

[3] Chang, H.D., Wang, J.F. and Suen, H.M. (1993), "Dynamic handwritten Chinese signature verification", *Proc. of ICDAR'93*, pp. 258 – 261.

[4] Cover, T.M. and Van Campenhout, J.M. (1977), "On the possible orderings in the measurement selection problem," *IEEE Trans. Systems, Man, and Cybernetics*, Vol. 7, No. 9, pp. 657-661.

[5] Crane, H.D. and Ostrem, J.S. (1983), "Automatic Signature Verification Using a Three-Axis Force-Sensitive Pen," *IEEE Trans. Systems, Man, and Cybernetics*, Vol. 13, No. 3, pp. 329-337, 1983.

[6] Fung, G.S.K., Liu, J.N.K. and Lau, R.W.H. (1996), "Feature Selection in Automatic Signature Verification Based on Genetic Algorithms," *Progress in Neural Information Processing*, Amari, et al., Eds., Springer-Verlag, pp. 811-815.

[7] Fung, G.S.K., Liu, J.N.K., Chan, K.H. and Lau, R.W.H. (1997), "Fuzzy genetic algorithm approach to feature selection problem", *Proc. of 6th IEEE International Conference on Fuzzy Systems*, July 1-5, 1997, Barcelona, pp. 441-446.

[8] Fung, G.S.K., Lau, R.W.H. and Liu, J.N.K (1997), "A signature based password authentication method," in *IEEE International Conference on Systems, Man, and Cybernetics (SMC'97)*, October 12-15, 1997, Orlando, Florida, U.S.A., pp. 631-636.

[9] Hamamoto, Y., Uchimura, S., Matsunra, Y., Kanaoka, T. and Tomita, S. (1990), "Evaluation of the Branch and Bound Algorithm for Feature Selection," *Pattern Recognition Letters*, Vol. 11, pp. 453-456.

[10] Herbst, N.M. and Liu, C.N. (1982), "Automatic signature verification," *Computer Analysis and Perception*, C.Y. Suen and R. De Mori, Eds., CRC Press, pp. 83-105.

[11] Huang, L.K. and Wang, M.J.J. (1996), "Efficient shape matching through model-based shape recognition," *Pattern Recognition*, Vol. 29, No. 2, pp. 207-216.

[12] Jain A. and Zongker, D. (1997), "Feature Selection: Evaluation, Application, and Small Sample Performance," *IEEE Trans. Pattern Analysis and Machine Intelligence*, Vol. 19, No. 2, pp. 153-158.

[13] Keller, J.M., Gray, M.R. and Givens, J.A. Jr. (1985), "A Fuzzy K-Nearest Neighbor Algorithm," *IEEE Trans. Systems, Man, and Cybernetics*, Vol. SMC-15, No. 4, pp. 580-585.

[14] Kittler, J. (1978), "Feature Set Search Algorithms," *Pattern Recognition and Signal Processing*, C.H. Chen, Ed., pp. 41-60.

[15] Klir, G.J. and Yuan, B. (1995), "Fuzzy Systems and Genetic Algorithms," *Fuzzy Sets and Fuzzy Logic: Theory and Applications*, pp. 452-454, Prentice Hall International.

[16] Lee, L.L., Berger, T. and Aviczer, E. (1996), "Reliable On-Line Human Signature Verification Systems," *IEEE Trans. Pattern Analysis and Machine Intelligence*, Vol. 18, No. 6, pp. 643-647.

[17] Lee, S. and Pan, J.C. (1992), "Offline tracing and representation of signatures," *IEEE Trans. Systems, Man, and Cybernetics*, Vol. 22, No. 4, pp. 755-771.

[18] Leclerc, F. and Plamondon, R. (1994), "Automatic Signature Verification: The State of Art – 1989-1993", *International Journal of Pattern Recognition and Artificial Intelligence*, Vol. 8, No. 3, pp. 643-660.

[19] Liu, J., Gross, S. and Murray, G. (1994), "Similarity comparison and analysis of sequential data," *Proc. of International Conference on Expert Systems for Development*, Asian Institute of Technology, Bangkok, Thailand, March 28-31, 1994, pp. 138-143.

[20] Liu, J.N.K. and Lee, R.S.T. (1997), "Invariant character recognition in Dynamic Link Architecture," in *Knowledge and Data Engineering Exchange Workshop (KDEX'97) of the 9th IEEE Tools with Artificial Intelligence Conference*, November 3, 1997, Newport Beach, California, U.S.A., pp. 188-195.

[21] Liu, J.N.K. and Lee, R.S.T. (1998), "Invariant handwritten Chinese character recognition," *Proc. of ICONIP/JNNS'98*, October 21-23, 1998, Kitakyushu, Japan, pp. 275-278.

[22] Mao, J., Mohiuddin, K. and Jain, A.K. (1994), "Parsimonious Network Design and Feature Selection through Node Pruning," *Proc. of 12th International Conference on Pattern Recognition*, pp. 622-624.

[23] Mitra, S., Pal, S.K. and Kundu, M.K. (1994), "Fingerprint classification using a fuzzy multilayer perceptron," *Neural Computing and Application*, Vol. 2, No. 4, pp. 227-233.

[24] Narendra, P.M. and Fukunaga, K. (1977), "A Branch and Bound Algorithm for Feature Subset Selection," *IEEE Trans. Computers*, Vol. 26, No. 9, pp. 917-922.

[25] Pudil, P., Novovicova, J. and Kittler, J. (1994), "Floating Search Methods in Feature Selection," *Pattern Recognition Letters*, Vol. 15, pp. 1119-1125.

[26] Sanchez, E. (1994), "Soft Computing Perspectives," *Proc. of 24th International Symposium on Multiple-Valued Logic*, pp. 276-281, 1994.

[27] Sato, Y. and Kogure, K. (1982), "On-line signature verification based on shape, motion and handwritten pressure," *Proc. of 6th International Conference on Pattern Recognition*, pp. 823 – 826.

[28] Siedlecki, W. and Sklansky, J. (1988), "On Automatic Feature Selection," *International Journal of Pattern Recognition and Artificial Intelligence*, Vol. 2, No. 2, pp. 197-220.

[29] Wells, G.L. and Loftus, E.F. (Eds.) (1984), *Eyewitness Testimony –Psychological Perspectives*, Cambridge University Press.

[30] Yasuhara, M. and Oka, M. (1977), "Signature verification experiment based on nonlinear time alignment: A feasibility study," *IEEE Transactions on Systems, Man and Cybernetics*, SMC-7, pp. 212-216.

[31] Yu, B. and Yuan, B. (1993), "A More Efficient Branch and Bound Algorithm for Feature Selection," *Pattern Recognition*, Vol. 26, No. 6, pp. 883-889.

Chapter 7:

The Application of a Generic Neural Network to Handwritten Digit Classification

THE APPLICATION OF A GENERIC NEURAL NETWORK TO HANDWRITTEN DIGIT CLASSIFICATION

D.S. Banarse and **A.W.G. Duller**
School of Electronic Engineering & Computer Systems
University of Wales, Bangor
Dean Street, Bangor, Gwynedd LL57 1UT, United Kingdom
email: A.Duller@sees.bangor.ac.uk

This chapter investigates the application of a self-organizing neural network to handwritten character recognition. While many approaches have been suggested for this application the proposed method has the advantage of being a more generic solution which is capable of being used in a range of image recognition tasks with the need for very little application-specific knowledge to be provided to the network. Results are presented using the NIST database as the source for the characters.

1 Introduction

This chapter describes the application of the PARADISE (PAttern Recognition Architecture for Deformation Invariant Shape Encoding)[1] neural network to the recognition of handwritten characters from the NIST database. The PARADISE network was designed to be a generic recognition system with the solution of many recognition problems requiring a minimum amount of application-specific knowledge to be provided. It has been used to recognize a range of object types including hand postures [2], arbitrary shapes [1] and human faces [3]. The deformable templates that are constructed by the network are based on feature detectors which to some degree are problem dependent. However, it has been found that a tunable zero integral Gabor filter, effectively identifying edges at a given scale, provides a generally usable solution. In addition to the Gabor filter, this chapter describes the use of oriented Gaussian filters as an alternative form of feature detector.

A great deal of research has been directed towards the field of character recognition [4], [5], [6], [7]. Biologically inspired character recognition approaches [8], [9], [10] show much promise by providing a powerful method of incorporating robustness to character deformations. This is achieved by employing architectural aspects of the mammalian vision system, namely, a hierarchical structure involving the recognition of increasingly more complex patterns as signals progress from stage to stage through the network. One of the most renowned biologically inspired character recognition networks is the neocognitron [9], which is based on a hierarchy of processing stages providing the system with rotation, translation, scale and deformation invariance. Practical use of the neocognitron is difficult due to its large number of connections and complex self-organizing nature. The architecture of PARADISE has been designed to exploit the deformation invariant properties of the neocognitron while minimizing the complexity of the network.

2   The PARADISE Network

A brief overview of the PARADISE network is now given. For more details see [1], [11]. To accomplish object recognition, it constructs templates for each class based on "component patterns" which are small local patterns of features extracted from the images presented. Several component patterns are linked to a single cell in the classification layer to represent a class. The network has a three layer architecture:

1. The Feature Extraction (FE) layer.

2. The Pattern Detection (PD) layer.

3. The Classification layer.

2.1   The Feature Extraction Layer

The FE layer consists of a single layer of FE planes. Each plane extracts the "strength" of a given type of feature at all possible points in the input image. For this work, two FE planes were used with either the Gabor filters or the Gaussian filters, oriented vertically on one plane and horizontally on the other.

2.2   The Pattern Detection Layer

This layer builds up a library of "component patterns" which are small local patterns of features produced by the FE layer. These localized component patterns are used as the basis for forming class templates. Owing to their relatively small size, the component patterns can often be reused to represent similar parts of many objects. New component patterns are automatically generated during learning and the network architecture makes them inherently robust to small translations and deformations. The radius of the component patterns is determined by the parameter λ and the amount of translational tolerance that is allowed for each component pattern is γ.

2.3   The Classification Layer

Each cell in this layer encodes a template for a class by creating links to the pattern detection layer. Thus each class template is composed of a number of component patterns connected to a single classification cell. When subsequent objects are presented to the network, the existing classes are examined to see if a sufficiently good representation already exists; if not, a new class is created. When an object is recognized, the component patterns of the "winning" template are adjusted to better represent the component patterns of the presented object, the amount of adjustment being controlled by $0 \leq \delta \leq 1$. This adjustment process is known as "weight update" and is discussed in a later section with respect to Gabor feature extraction.

2.4   Network Parameters

A number of parameters can be set to control the type of recognition performed by the network. The majority can be set using heuristics [11] and generally stay fixed for a given application. Once these are set, the "classification threshold" parameter, α, is used to determine the degree of match that is required between the input object and the internal template model. While the classification threshold greatly affects the response of the network, it has been shown [1] that in terms of the qualitative results it is only important to get the value in the "right area", i.e., the changes in network behavior are gradual with changing threshold.

2.5   Network Dynamics

The architecture of the network provides a built-in tolerance to deformation of object images. Relatively few examples of objects are needed for training because the network architecture automatically generalizes to recognize similar shapes. This results in most applications requiring only a small number of passes through the data set to achieve learning. Using on-line learning, the network can be used to cluster a data set; the validity of the clusters can be evaluated through examining the quality of the clusters after the number of classes within the network has stabilized.

3   Applying the Network

3.1   The Image Data Set

The handwritten numerals used were 32×32 pixel binary images from the NIST database. The whole set of images as used, making a training set of 3471 examples written by 49 different authors. An example of the variation within the different digits in the database is shown in Figure 1.

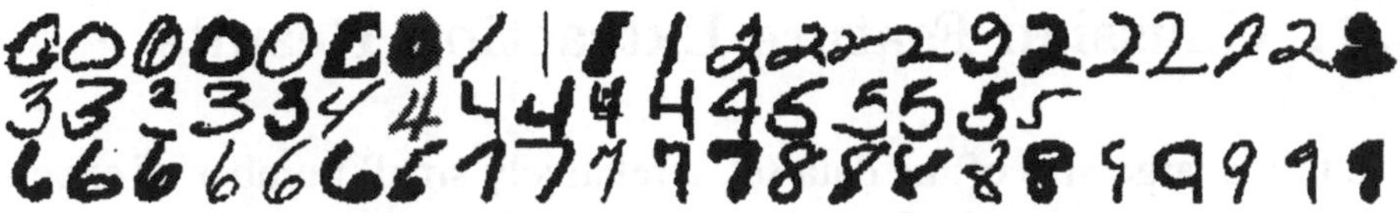

Figure 1. Examples of the variation of digits in the NIST database.

3.2   Application-Specific Details

In order to use the network for a new recognition application, the most important aspect is to determine the type of feature extraction that is best suited to the objects that are to be recognized. Two methods of feature extraction were chosen to be tested, Gabor filters and oriented Gaussian filters. The Gabor filters were tuned to be most sensitive to edges that span six pixels; higher frequency Gabors would detect two edges for a pen stroke, one edge on either side of the stroke. The oriented Gaussian filters were chosen since it was thought that edge detection could complicate the task, making the network overly sensitive to the width of the lines used in the images. The Gaussian filters provide a low pass filtering of the image at specified orientations. The oriented Gaussians are given by

$$\mathbf{G_{FE}} = \exp\left(-\frac{(x\cos\theta + y\sin\theta)^2}{2\sigma_x^2} + \frac{(-x\sin\theta + y\cos\theta)^2}{2\sigma_y^2}\right) \tag{1}$$

where σ_x and σ_y control the width of the Gaussian filter in the x and y directions, respectively, and θ determines the orientation. The width of the filter is designed such that the Gaussian values tail off to 0.01 at the corners of the filter mask.

The network was employed in a self-organizing manner, with the digits presented in a random order. The sequence of digits was repeatedly presented to the network until the number of classes stabilized. In all of the experiments presented here, at most 4 passes through the data set were required. At this point, the final classification of each digit was obtained through a final presentation cycle. Each class was assigned a "metaclass" determined by the most frequently represented digit in that class. This allowed the number of correct classifications to be calculated with respect to the metaclasses.

3.3   Gaussian Feature Extraction Results

Due to the large size of the data set, a relatively small number of experiments were performed. Parameters were individually chosen for each experiment guided by previous results and heuristics. The basic network parameters that remained the same throughout the experiments are listed below:

$\sigma_x = 2$ σ_y was then adjusted the shape of the Gaussian.

$\gamma = 6$ the translational tolerance was fixed at a level that appeared reasonable for the digit images.

Table 1 shows the results from three experiments with a range of numbers of classes having been created. For reference, the "variable" parameters are shown on the right-hand side of the table.

Table 1. Results using oriented Gaussian feature extraction.

Classes	Classes per metaclass	% correct	α	σ_y	λ
55	5.5	71.2	0.5	4	6
78	7.8	67.3	0.7	4	5
125	12.5	81.5	0.75	6	8

3.4   Gabor Feature Extraction Results

The experiments involving Gabor feature extraction were used to test the effectiveness of the Gabor filters and also the effect of weight update on class creation. The only parameter that was varied between experiments was δ which controls the adjustment of the component patterns. $\delta = 0$ gives no weight update. The rest of the network parameters remained fixed and are listed below:

$\alpha = 0.25$ was chosen using the heuristics.

$\lambda = 8$ the radius of the component patterns was set to a size that constitute small line segments of characters.

$\gamma = 6$ the translational tolerance was fixed at a level that appeared reasonable for the digit images.

The results are shown in Table 2. For comparison, results are also shown for $\alpha = 0.42$ and $\delta = 0$, i.e., no weight update. The heuristically chosen value of $\alpha = 0.25$ has produced a solution with a remarkably small number of classes per metaclass and this has caused a relatively poor performance in terms of the percentage correct due to the wide variation in some of the digits. Hence, the inclusion of the results for the higher value of α, which give far better discrimination at the cost of significantly more classes. With these handwritten digits weight update hinders good performance. This is due to the large variation in the characters in the data set which leads to component patterns that learn to detect patterns that are too general. The final classifications for the digits in the experiment where $\alpha = 0.42$ are shown in Figures 2, 3 and 4, where each row represents a separate class. For each row the images which are most confidently recognized are on the left. On the extreme left of each row is the class number and in brackets the number of images that were recognized by that class. Many of the classes are too large to be displayed in their entirety; so for these, three sections have been extracted, the extreme left, the middle and the extreme right portions of the row of images.

It can be seen that many of the classes recognize relatively few digits; some only recognize the single image upon which they were trained. It is often found that these classes are created around very distorted images (e.g., class 26) that do not clearly resemble the digit they are meant to represent. Examples of such images are included in Figure 1. The network is sensitive to character styles, for example, the different forms of writing a '7' (Classes 27, 34, 41, 57, 60, 66 and 87). It is also interesting to note the most common types of misclassification, for example, the numbers '4', '7' and '9' are often confused. Particular styles of these digits look very similar, e.g., the first few examples of class 71 in Figure 4. Similarly, the digits '3', '5' and '8' are often confused.

Table 2. Results using Gabor feature extraction and weight update.

Classes	Classes per metaclass	% correct	δ	α
18	1.8	54.9	0.1	0.25
17	1.7	54.5	0.01	0.25
22	2.2	67.0	0	0.25
91	9.1	87.9	0	0.42

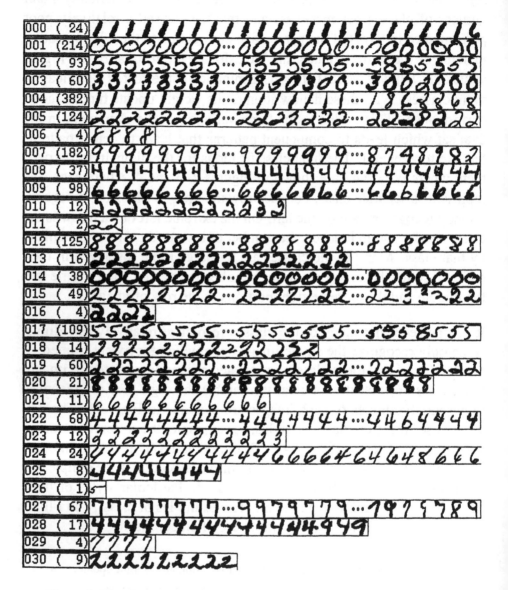

Figure 2. The classes generated using Gabor feature extraction, classes 0-30.

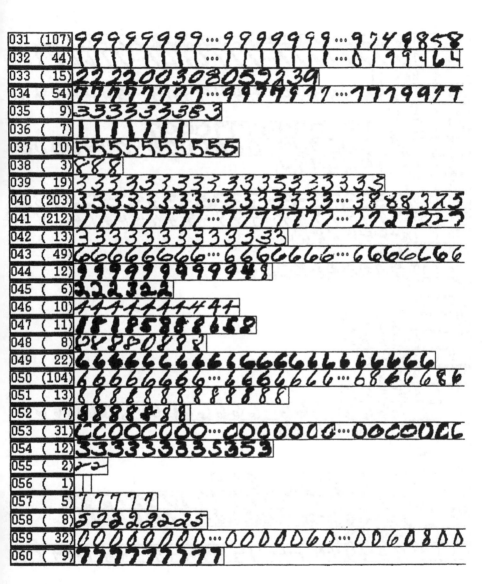

Figure 3. The classes generated using Gabor feature extraction, classes 31-60.

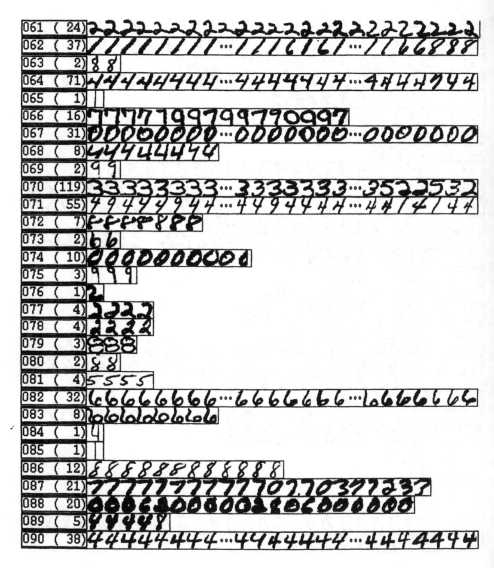

Figure 4. The classes generated using Gabor feature extraction, classes 61-90.

The network is also sensitive to the writing implements used as they can change the pen stroke thickness to a large degree. To overcome this, a feature extraction that is insensitive to the line thickness or a preprocessing technique to skeletonize the characters could be used. However, these would involve their own problems and limitations. In many cases, it would be more desirable to use the Gabor feature extraction and have a separate class for the different styles, linking them together if necessary at a higher level outside the network.

3.5 Analysis of Component Patterns

To provide an insight into the workings of PARADISE, this section shows what portions of the object are used to create the class templates. Figure 5 shows a digit '3' that was used to create one of the classes in the exper-

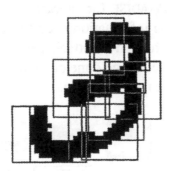

Figure 5. The placement of component patterns for a class template.

iments. The squares mark the areas of the image that correspond to the component patterns used to form the class template, although the actual component patterns learned are the outputs from the FE planes at these points. The translational tolerance which determines the area over which each component pattern can be detected around its original position was slightly smaller than the component pattern areas.

The component patterns correspond to overlapping areas of the image. The individual component patterns are more clearly illustrated in Figure 6 where the circles represent the centres of the component pattern areas and the patterns on the periphery represent the image segments that correspond to each of the component patterns. Similar objects are likely to be composed of similar component patterns in similar locations and even dissimilar shapes may share some similar component patterns.

4 Summary

The general-purpose P̲ARADISE network has been shown to be a useful tool for this particular application. With very few application specific changes to the network, good performance is achievable. The results show that by using on-line learning in a self-organized manner, the network can successfully generate new class templates for novel images, partitioning the data set into separate classes. The quality (percentage of correct classifications) of the classes is traded against the number of classes, controlled by the classification threshold. It is recognized that to extend the accuracy of P̲ARADISE for character recognition, mechanisms for incorporating *a priori* information need to be sought. This may be achieved by enabling a human to attach particular importance to key component patterns; this could be done on an incremental basis throughout the early stages of the network's life.

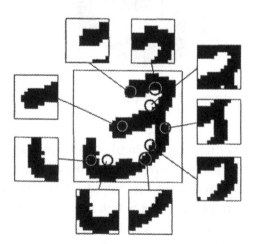

Figure 6. The component patterns for a digit class.

The percentage of correct classifications is higher when the Gabor filters are used for feature extraction. For a similar level of performance, considerably fewer classes were needed with Gabor filtering than oriented Gaussians; for example, with the oriented Gaussian filters, 78 classes were required to give comparable performance to a solution requiring only 22 classes when using Gabor filters. A relatively large number of classes were required (in comparison to the number of real-world classes) to attain a reasonably high level of performance ($\approx$ 9 classes per metaclass). However, this is not surprising considering the variable number of styles and writing implements displayed and the variable quality of some of the digits in the data set.

References

[1] Banarse, D.S. and Duller, A.W.G. (1997), "Vision Experiments with Neural Deformable Template Matching," *Neural Processing Letters*, 5(2):111–119.

[2] Banarse, D.S. and Duller, A.W.G. (1996), "Deformation invariant pattern classification for hand gesture recognition," *Proceedings of IEEE International Conference on Neural Networks*, pp. 1812–1817.

[3] Banarse, D.S. (1997), *A Generic Neural Network Architecture for Deformation Invariant Object Recognition*. Ph.D. thesis, University of Wales, Bangor.

[4] Casey, R.G. and Lecolinet, E. (1996), "A survey of methods and strategies in character segmentation," *IEEE Transactions on Pattern Analysis and Machine Intelligence*, 18(7):690–706.

[5] Bailey, R.R. and Srinath, M. (1996), "Orthogonal moment features for use with parametric and non-parametric classifiers," *IEEE Transactions on Pattern Analysis and Machine Intelligence*, 18(4):389–399.

[6] Liou, C.-Y. and Yang, H.-C. (1996), "Handprinted character recognition based on spatial topology distance measurement," *IEEE Transactions on Pattern Analysis and Machine Intelligence*, 18(9):941–945.

[7] Wilfong, G., Sinden, F., and Ruedisueli, L. (1996), "On-line recognition of handwritten symbols," *IEEE Transactions on Pattern Analysis and Machine Intelligence*, 18(9):935–940.

[8] Shustorovich, A. and Thrasher, C.W. (1996), "Neural network positioning and classification of handwritten characters," *Neural Networks*, 9(4):685–693.

[9] Fukushima, K. and Imagawa, T. (1993), "Recognition and segmentation of connected characters with selective attention," *Neural Networks*, 6:33–41.

[10] Hussain, B. and Kabuka, M.R. (1994), "A novel feature recognition neural network and its application to character recognition," *IEEE Transactions on Pattern Analysis and Machine Intelligence*, 16(1):98–106.

[11] Banarse, D.S. and Duller, A.W.G. (1997), "Deformation invariant visual object recognition: Experiments with a self-organising neural architecture," *Neural Computing and Applications*, 6(2):79–90.

Chapter 8:

High-Speed Recognition of Handwritten Amounts on Italian Checks

HIGH-SPEED RECOGNITION OF HANDWRITTEN AMOUNTS ON ITALIAN CHECKS

B. Lazzerini
Dipartimento di Ingegneria della Informazione
Università degli Studi di Pisa
Via Diotisalvi, 2, 56126 Pisa, Italy

L.M. Reyneri, F. Gregoretti
Dipartimento di Elettronica
Politecnico di Torino
C.so Duca degli Abruzzi, 24, 10129 Torino, Italy

A. Mariani
Macs Tech, s.r.l.
Via S. Paolo, 11, 56125 Pisa, Italy

In this chapter we describe an Italian bank check processing system. The system is made up of several processing modules, including those for data acquisition, image preprocessing, character center detection, character recognition, courtesy amount recognition, and legal amount recognition. The system separately reads and recognizes (at the character level) the courtesy and legal amounts. Then the two streams of information are sent to a context analysis subsystem, which exploits the mutual redundancy in the courtesy and legal amounts. The output of the system is a list of possible amounts, which are sorted according to a decreasing recognition confidence. The system has an accuracy of 53% and can process up to three bank checks per second. The proposed system is partially implemented on an ad-hoc VLSI chip containing an array of dedicated processors tailored to the application, and tightly interconnected to a host personal computer.

1 Introduction

This work describes HACRE, a high-speed system for real-time automatic recognition of handwritten Italian checks. The aim of our work was to recognize real-world checks, where handwriting is assumed to be un-boxed and usually unsegmented, so that characters in a word may touch or even overlap.

The amount on Italian checks consists of only one word, thus no phrase segmentation is required. The amount is written twice: the *legal amount* (namely, the literal one), and the *courtesy amount* (namely, the numerical one). The two fields are placed in well-known areas of the check, and an approximate localization of these two areas can be obtained from the information contained in the *code-line* printed at the bottom of the check (assuming that each bank has its own check layout, as is usually the case).

Experimental results have shown that an acceptable recognition performance for a given application can be achieved only by using redundancy, namely, contextual information [1], [2]. In the specific application, redundancy is present both in the exact correspondence that exists between the legal and the courtesy amounts, and in the very limited size of the dictionary to be used (combinations of about 30 words, for Italian checks); also, it is interesting to note that only 15 out of the 26 letters of the alphabet are used in such a dictionary.

Furthermore, the amount of computations required for a reliable recognition of handwritten text is very high; therefore, real-time constraints can be satisfied only by using very powerful processors. We have therefore decided to develop a dedicated processing architecture, to cope with tight cost constraints and size requirements, and to tailor the recognition algorithms to the architecture capabilities, to get the highest performance from the hardware.

The proposed system integrates the five subsystems shown in Figure 1:

- a mechanical and optical *scanner*, to acquire a bit-map image of the check;
- an *image preprocessor*, for preliminary filtering, scaling, and thresholding of the image;

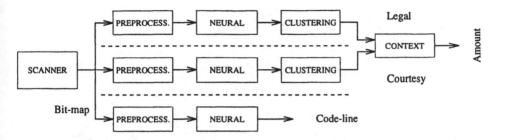

Figure 1. Block diagram of HACRE.

- a *neural subsystem,* based on an ensemble of neural networks and feature-based recognizers, which detect characters and provide hypotheses of recognition for each detected character;
- a *clustering subsystem* which improves the performance of the centering detector of the *neural subsystem*;
- a *context analysis subsystem* based on a lexical and syntactic analyzer.

Legal and courtesy amounts are independently preprocessed and recognized up to the character level; the two streams of information are then sent to the common *context analysis subsystem,* which exploits all the mutual redundancy.

In practice, the *code-line* field located at the bottom of the checks is also scanned and processed to obtain general information about the bank and, indirectly, about check layout. Recognition of this field is easy, as it is printed on a light background and its graphical quality is usually very high; therefore, it is not discussed here.

The *neural* and *clustering subsystems* together carry out a prerecognition of the individual characters, based on an *integrated segmentation and recognition* technique [3]. In correspondence to each character of a handwritten word, the *neural subsystem* of HACRE produces a list of "candidate" characters, instead of just one as in other recognizers.

The error rate of the *neural* and *clustering subsystems* alone is high, but this is strongly reduced by the *context analysis subsystem,* which combines the candidate characters and, guided by the mutual redundancy present in the legal and courtesy amounts, produces hypotheses about

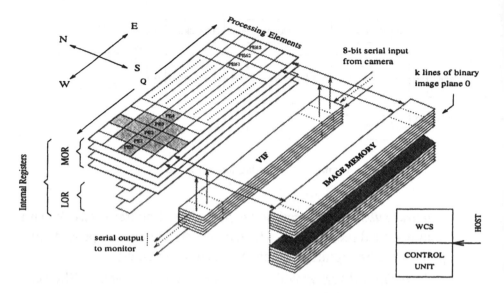

Figure 2. General architecture of the processor array.

the amount so as to correct errors made by the *neural* and *clustering* *subsystems*. At the end, the system generates a list of possible amounts, sorted according to a *recognition confidence index*.

1.1   Hardware Platform

The HACRE system was developed to run on a specific hardware platform (PAPRICA-3) [4], purposely designed at the Politecnico di Torino, and tightly interfaced to a host personal computer (for instance, a Pentium). The former implements the *image preprocessor* and the *neural subsystem*, while the host computer implements the *clustering* and the *context analysis subsystems*.

As shown in Figure 2, the kernel of PAPRICA-3 is a linear array of 64 identical 1-bit Processing Elements (**PE**s) connected to an image memory via a bidirectional 64-bit wide bus. The image memory is organized into addressable words whose length matches the size of the processor array. Each word contains data relative to one binary pixel plane (also called *layer*) of one line (64 bits wide) of an image. A single clock cycle is needed to load an entire line of data into the PE's internal registers.

Data can be transferred into the internal registers of each PE, processed and explicitly stored back into memory according to a RISC-like processing paradigm.

Each PE is composed of a Register File and a 1-bit Execution Unit, and processes one pixel of each line. The core of the instruction set is based on *morphological operators* [5]: the result of an operation depends, for each processor, on the value of pixels in a given neighborhood (5×5, as sketched by the grey squares in Figure 2). Data from E, EE, W and WW directions (where EE and WW denote pixels two bits apart in E and W directions) may be obtained by direct connection with neighboring PEs, while all other directions correspond to data of previous lines (N, NE, NW, NN) or subsequent lines (S, SE, SW, SS).

To obtain the outlined neighborhood, a number of internal registers (16 per each PE), called *Morphological Registers* (MOR), have a structure which is more complex than that of a simple memory cell, and are actually composed of five 1-bit cells with a S→N shift register connection. When a load operation from memory is performed, all data are shifted northward by one position and the south-most position is taken by the new line from memory. In this way, data from a 5×5 neighborhood are available inside the array for each PE, at the expense of a two-line latency. The remaining internal registers (48 per each PE), called *Logical Registers* (LOR), are only 1-bit wide.

The instruction set includes logical and algebraic operations which act either on a Logical Register or on the central bit of a Morphological Register.

An important characteristic of the system is the integration of a serial-to-parallel I/O device, called *Video InterFace (VIF)*, which can be connected to the linear CCD array for direct image input (and optionally to a monitor for direct image output). The interface is composed of two 8-bit, 64-stage shift registers which serially and asynchronously load/store a new line of the input/output image during the processing of the previous/next line.

Two inter-processor communication mechanisms are also available to exchange information among PEs which are not directly connected (not shown in figure). The first one is a *Status Evaluation Network*, to which each processor sends the contents of one of its registers, and which provides a *Status Word* divided into two subfields. The first one is composed of two global flags, named *SET* and *RESET*, which are true when the contents of the specified layer are all '1's or all '0's, respectively. The second one is the *COUNT* field which is set equal to the number of PEs in which the content of the specified register is '1'.

This global information may be accumulated and stored in the *Status Register File* and used for further processing or to conditionally modify the program flow. Status Registers can also be read by the host processor.

The second communication mechanism is an *Inter-processor Communication Network*, which allows global and multiple communications among clusters of PEs. The topology of the communication network may be varied at run-time.

A Host Interface allows the host processor (a Pentium, or any other PC) to access the Program and Image memories and some configuration registers. The access is through a conventional 32-bit data bus with associated address and control lines.

The HACRE system is composed of one PAPRICA-3 chip plus a fast static RAM providing a 128K×64 bit Image Memory, 256 Status Registers, a host Pentium processor, and an interface with the CCD imager. PAPRICA-3 has been designed to operate with a clock frequency of up to 100MHz.

2   Image Preprocessor

The *image preprocessor* of HACRE, which is shown in Figure 3, consists of the following blocks:

1. The WINDOW EXTRACTOR acquires the image from the SCANNER, at a resolution of approximately 200 dpi, grayscale, 16 levels. The scanner is an 876-pixel CCD line camera mechanically scanned over the image, from right to left (for practical reasons), at a speed of 2m/s (namely, about 700 characters/s).

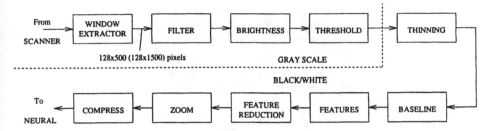

Figure 3. Block diagram of the image preprocessor.

Image size is about 876×1500 pixels, although characters are constrained in two smaller areas of known coordinates. These areas (128×500 and 128×1500 pixels) are extracted from the SCANNER by means of an ad-hoc CCD controller. Figure 4.a shows an "easy" example.

2. The FILTER block computes a simple low-pass filter with a 3×3 pixel kernel. The result is shown in Figure 4.b.

3. The BRIGHTNESS block compensates for the nonuniform detector sensitivity and paper color. A pixel-wise adaptive algorithm shifts the white level to a predefined value. See Figure 4.c.

4. The THRESHOLD block converts the grayscale image, after brightness compensation, into a B/W image; the brightness of each pixel is compared against an adaptive threshold which is a function of both the white and the black levels (see Figure 4.d).

   From this point onward, only the B/W image is processed. The THRESHOLD block also filters and partially removes the spot noise from the image, as shown in Figure 4.e.

5. The THINNING block reduces the width of all the strokes to 1 pixel, as shown in Figure 4.f. Thinning is a morphological operator [5] which reduces the width of lines, while preserving stroke connectivity.

6. The BASELINE block detects the *baseline* of the handwritten text, which is a horizontal stripe intersecting the text at a known position, as shown in Figure 4.g (left side). Unlike the other preprocessing steps, the baseline cannot be detected only by means of *local* algorithms (namely, algorithms with a limited neighborhood), as it is a *global* parameter of the entire image.

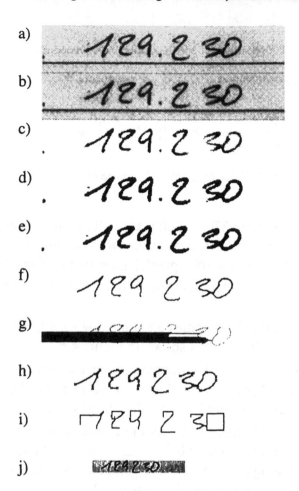

Figure 4. Preprocessing steps of handwritten images: a) original image, 200 dpi, 16 gray levels; b) low-pass filtered; c) compensated for brightness; d) thresholded; e) spot noise removal; f) thinned, after 6 steps; g) baseline (at the left side); h) features detection (features are tagged by small crosses); i) feature reduction; j) compressed.

The baseline is computed throughout the scan of the whole image and, therefore, is available only at the end of the image. The baseline is used to reduce the height of the processed image, from 128 pixels (minimum size to guarantee catching the whole manuscript, independent of its position, within a ±5mm tolerance boundary) to 64 pixels (to match the processor size).

7. The FEATURES block detects and extracts from the image a set of 12 *stroke features*, which are helpful for further character recog-

nition. This block detects the four *left*, *right*, *up* and *down con-cavities*, and the *terminal strokes* in the eight main directions (see Figure 4.h).

Features are helpful both for center localization (see Section 3.1) and for character recognition (see Section 3.3), as their types and positions almost univocally identify the character to be recognized. Features are not used alone, but together with the neural recognizer to improve recognition reliability.

8. The FEATURE REDUCTION block reduces the number of features, by removing both redundant and useless features. Examples of redundant features can be seen in the vertical stroke of digit "9". The result is shown in Figure 4.i.

9. The ZOOM block evaluates the *vertical size* of the manuscript and scales it in size to approximately 25-30 pixels. Vertical size is an approximate measure of calligraphy size and is defined as the minimum height of a rectangle which contains about 90% of the strokes (the percentage may vary for literals).

10. The COMPRESS block further reduces image size, approximately by a factor of 2, by means of an ad-hoc topological transformation which does not preserve image shape, although it preserves its connectivity, as shown in Figure 4.j. This transformation depends on the features detected, as areas containing fewer features are compressed more than others.

At this point, after all the preprocessing steps, the B/W image is ready for the following recognition steps, as described in the next sections. The image has been reduced both in size (down to $14 \times 18 = 252$ or $12 \times 21 = 252$ pixels, for the courtesy and legal amounts, respectively), in numbers of gray levels (2), and in stroke thickness (1 pixel), and noise was removed. Table 3 lists the execution time of all the individual blocks.

At the end, each character fits into 256 bits, which are then reorganized as four adjacent processor words (4×64 bits) in order to optimize the performance of the hardware implementation.

3   Neural Subsystem

The second subsystem of HACRE is a hybrid neural network recognizer combined with a feature-based recognizer. Many existing handwriting recognizers [1]-[4], [6] require two consecutive steps, namely *character segmentation* and *character recognition*.

Unfortunately, segmentation is quite a difficult task [7], especially for handwritten texts, as there is no clear separation between consecutive characters, as shown in Figure 5. Furthermore, the same sequence of strokes can often be interpreted in different ways. For instance, the strokes shown in Figure 5.b can either be interpreted (also by humans) as "n mi", "nnn", "nnu", "n ini", "vvv", "vwi", etc.

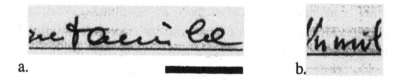

a.   b.

Figure 5. Problems arising in handwriting segmentation: a) difficulties in segmentation; b) ambiguities and pseudo-characters.

An erroneous segmentation process sends the wrong data to the character recognition step. It adds a number of unwanted artifacts and increases the recognition error rate.

We therefore decided to use an *integrated segmentation and recognition* approach [3], in which segmentation is tightly integrated with recognition. In practice, we first detect character *centers* (see Figure 6); then we apply the recognition algorithms (see Section 3.3) only to the windows centered around them, without segmenting and isolating characters.

3.1   Centering Detection

The CENTERING DETECTOR scans the preprocessed and compressed image from right to left and extracts a *sliding window* of 32 × 64 pixels, namely, one window per each preprocessed line. Note that windows without strokes are immediately skipped, as they contain no useful in-

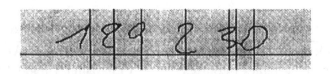

Figure 6. Character centers found by the CENTERING DETECTOR.

formation. Each window is associated with its *window coordinate x*, which is the distance in pixels of the window's geometrical center from the right-hand side of the check.

The CENTERING DETECTOR then computes, for each new window coordinate, a *centering function* $C_{\mathcal{F}}(x)$, which depends on several image characteristics, as described below. The centering function contains local maxima, which roughly coincide with and, therefore, detect the locations of character centers.

The CENTERING DETECTOR computes $C_{\mathcal{F}}(x)$ according to the relative position, type and quantity of the features detected by the FEATURES block. The centering function also depends on a weighted average of some geometrical characteristics of the characters, such as: i) text height at the center of the sliding window, ii) stroke curvature in correspondence with the top-most feature, iii) distance between adjacent vertical strokes, etc.

Figure 6 shows the results of the center detector. Solid lines are drawn in correspondence to the peaks of the centering function $C_{\mathcal{F}}(x)$. Sometimes each character center is localized at more than one position, but the *clustering subsystem* easily filters and removes the redundant ones, as described in Section 4. Note also that the character's "center" is not necessarily at its geometrical center, but at a given predefined position, unique to each character.

As a result, the CENTERING DETECTOR delivers the following values to the *context analysis subsystem*, for the kth peak of $C_{\mathcal{F}}(x)$:

1. the window coordinate x_k of the peak;
2. the *centering confidence* $C_{\mathcal{F}}(x_k)$.

Table 1. A few pseudo-characters used by HACRE for legal and courtesy amounts.

Pseudo character	Amount legal	Amount courtesy	Pseudo character	Amount legal	Amount courtesy
o	a,o,O	0	o	o,O	0
d	d	-	1	-	1
i	i,u,m,n	-	2	-	2
c	c,C	-	3	-	3
e	e,l	-	4	-	4
m	m,n	-	5	-	5
r	r	-	6	-	6
t	t	-	7	-	7
v	v,V	-	8	-	8
a	a	-	9	-	9

3.2   Pseudo-characters

An additional feature of HACRE is its ability to recognize a set of so-called *pseudo-characters*, which mostly coincide with traditional characters, except for some differences which were explicitly introduced to improve recognition performance.

The idea behind the concept of pseudo-characters results from the analysis of the problems arising in the identification of character centers. For instance, the ambiguous trace in Figure 5.b can be recognized more easily and with less ambiguity if it is recognized as a sequence of six pseudo-characters ⁊. Most "m"s and "n"s consist of the sequence of three (respectively, two) pseudo-characters ⁊.

It is obvious that, since characters are not recognized in their original form, words must be translated according to the chosen set of pseudo-characters. The dictionary used by the *context analysis subsystem* must then be preprocessed. For the example in Figure 5.b, the dictionary will contain the word "U ⁊⁊⁊⁊⁊⁊ l...", instead of "Un mil..." (with an appropriate coding of non-ASCII pseudo-characters). We have outlined 46 pseudo-characters, including the 10 digits, about 30 alphanumerics, plus 6 delimiters. Table 1 shows a few of them.

As the correspondence between characters and pseudo-characters is not univocal, each word in the dictionary produces more than one equivalent using pseudo-characters. For example, the letter "a" can be translated into either "o", or "ci", "cl". Pseudo-characters are also used by the CENTERING DETECTOR.

3.3   Character Recognizer

The CHARACTER RECOGNIZER recognizes each individual pseudo-character, using a hybrid approach, which mixes feature-based [8] and neural [9] recognizers.

Feature-based recognition is a multi-method process. First of all, features extracted by the FEATURES block are used to identify all easy-to-recognize pseudo-characters. For instance, most "0," "6," "9" digits (but not only these) are written well enough that a straightforward and fast analysis of the main features and strokes is sufficient to recognize those characters with a high accuracy.

As an example, character "0" is recognized when 4 connected concavities (1 left, 1 right, 1 up and 1 down) are detected, and the size of the character is higher than 50% of the text *vertical size*.

In the second method of feature-based recognition, the approximate area of the pseudo-character is split into 3×3 blocks. In each block we count how many features of each type are present; then we match these counts with a "prototype" for each pseudo-character (in practice, this is a hybrid feature-based and neural approach, as the prototypes are trained using neural algorithms).

In the other feature-based methods, features are combined in a number of different ways, together with other information extracted by the *image preprocessor*, such as text height and width.

Some characters are more difficult to recognize using only features; for instance, digits "4," "7" and some types of "1" can be more easily recognized using neural techniques. HACRE isolates all characters which have not been recognized using features (or have been recognized with insufficient confidence) and passes them to a neural network trained by means of an appropriate training set.

The CHARACTER RECOGNIZER is "triggered" for each pseudo-character center detected by the CENTERING DETECTOR. As shown in Table 3, the CHARACTER RECOGNIZER is by far the slowest piece of code, due to the large number of synaptic weights involved. Fortunately it is run at a relatively low rate, namely, every 15 lines on an average; therefore, its effects on computing time are limited.

The neural recognizer consists of the cascade of two neural networks:

1. For the courtesy amount, a two-layer WRBF network [10] with 252 inputs, 100 hidden and 20 output units, one for each pseudo-character to be recognized (digits 0-9, plus a few delimiters). For the legal amount, a two-layer WRBF network with 252 inputs, 230 hidden and 46 output units, associated with as many pseudo-characters.

   This network is initialized as described in [10], without any further training. The training set used to initialize the network was obtained by manually classifying a number of examples (about 1,000 for the courtesy amount and 3,000 for the legal amount).

2. A one-layer MLP [11], with 20 inputs and 20 outputs (or 46 and 46, respectively, for the legal amount), which is trained to improve the quality of the output of the neural recognizer. The network was trained using an adaptive delta rule [11].

3.4   Characterization of the Character Recognizer

For each center coordinate x_k, the output of the CHARACTER RECOGNIZER is a list

$$\mathcal{L}(x_k) = \{(c_1, e^1(x_k)), \dots, (c_j, e^j(x_k)), \dots\} \tag{1}$$

of *recognition confidences* $e^j(x_k)$ associated with each pseudo-character c_j (c_1="0", c_2="1", ..., c_{20}=✦, ...). The confidence $e^j(x_k)$ is a weighted average of the confidence degrees computed by the feature-based recognizers and the outputs of the neural network. The list is then sorted according to decreasing values of e^j.

The recognition confidences are then processed before delivery to the *context analysis subsystem*, which receives the *weighted recognition confidences* $\hat{e}^j(x_k)$:

$$\hat{e}^j(x_k) = \sum_{n=1}^{N_M} P(c_j|c_{i(n),n}) \cdot e^{i(n)}(x_k) \tag{2}$$

where $P(c_j|c_{i,n})$ is the probability that the CHARACTER RECOGNIZER has in input pseudo-character c_j conditioned by the fact that in output it recognizes pseudo-character c_i in the nth position of the sorted list $\mathcal{L}(x_k)$, while $i(n)$ is the index of the pseudo-character in the nth position of the sorted list $\mathcal{L}(x_k)$, and N_M is the number of pseudo-characters.

The conditional probability $P(c_j|c_{i,n})$ is calculated, by applying Bayes' theorem:

$$P(c_j|c_{i,n}) = \frac{P(c_{i,n}|c_j) \cdot P(c_j)}{\sum_{l=1}^{N_M} P(c_{i,n}|c_l) \cdot P(c_l)} \tag{3}$$

where $P(c_{i,n}|c_j)$ is the probability that the CHARACTER RECOGNIZER recognizes pseudo-character c_i in the nth position of the sorted list $\mathcal{L}(x_k)$, when there is pseudo-character c_j in input. Instead, $P(c_j)$ is the probability that pseudo-character c_j is present in input. In practice, $P(c_{i,n}|c_j)$ is obtained by making statistics of the outputs of the CHARACTER RECOGNIZER, when recognizing all the pseudo-characters of the training set. $P(c_j)$ coincides with the appearance frequency of c_j in the training set.

For each center x_k, the *neural subsystem* produces a tuple

$$\mathcal{T}(x_k) \triangleq \left\{ x_k, C_{\mathcal{F}}(x_k), \hat{\mathcal{L}}(x_k) \right\} \tag{4}$$

where $C_{\mathcal{F}}(x_k)$ is the centering confidence, while $\hat{\mathcal{L}}(x_k)$ is the *list of alternatives* defined as

$$\hat{\mathcal{L}}(x_k) \triangleq \left\{ (c_1, \hat{e}^1(x_k)), \dots, (c_j, \hat{e}^j(x_k)) \dots \right\} \tag{5}$$

The list is then sorted according to decreasing values of the weighted recognition confidence $\hat{e}^j$.

4   Clustering Subsystem

The *clustering subsystem* aims to group the centers detected by the CENTERING DETECTOR into as many *clusters* as there are pseudo-characters in the courtesy (or legal) amount.

To this aim, some statistics are calculated on the training set. They include: i) the average intra-cluster distance, ii) the average inter-cluster distance, iii) the average number of centers per cluster, iv) the average pseudo-character width, and v) the average recognition confidences of the first 2 pseudo-characters in the sorted lists of alternatives.

These statistics provide a useful means to characterize the behavior of the CENTERING DETECTOR in typical cases like close or touching pseudo-characters, well-detached pseudo-characters, and nonrevealed pseudo-characters.

For example, in correspondence to a well detached pseudo-character, the CENTERING DETECTOR typically detects 1 (possibly 2) centers; when the pseudo-character is correctly recognized, it is usually present with a high recognition confidence in the first or second position of $\hat{\mathcal{L}}(x_k)$, and the average inter-cluster distance is about 2.5 to 3 times greater than the average intra-cluster distance.

The detection of these characteristics in the CENTERING DETECTOR output during the recognition of a new check amount is a clue to the presence of a well-detached pseudo-character. The collected statistics are used by the *clustering subsystem* to group centers into clusters.

More precisely, our implementation phase can be summarised as

1. We build two *basic centering functions*, $f_f(x - x_k)$ and $f_r(x - x_k)$, for each center x_k. The two functions are associated with, respectively, the centering confidence $C_{\mathcal{F}}(x_k)$ and the highest weighted recognition confidence ($\max_j \hat{e}^j(x_k)$) of the pseudo-characters in the list $\hat{\mathcal{L}}(x_k)$.

Basic centering functions are triangular functions defined as

$$f_f(x - x_k) \triangleq \begin{cases} C_{\mathcal{F}}(x_k) \cdot (1 - \frac{2|x-x_k|}{\bar{\omega}}), & \text{for } 2|x - x_k| \leq \bar{\omega} \\ 0, & \text{otherwise} \end{cases} \tag{6}$$

where $\bar{\omega}$ is the average pseudo-character width. A similar expression is used for $f_r(x_k)$, by substituting $C_{\mathcal{F}}(x_k)$ with $(\max_j \hat{e}^j(x_k))$.

2. We define a *global centering function* which sums up the two basic functions and appears as a sequence of peaks (local maxima) corresponding to centers and valleys (local minima).

$$\mathcal{F}(x) \triangleq \sum_k (f_f(x - x_k) + f_r(x - x_k)) \tag{7}$$

In this way, very close centers corresponding to the same pseudo-character quite often give origin to one peak of the global centering function.

The peaks are then classified as either *isolated* or *nonisolated*. A peak is isolated if and only if either of the following conditions holds: i) its nearest peak is at a distance greater than or equal to the average pseudo-character width, ii) it is much higher than its closest peak. We assume that isolated peaks identify as many clusters.

3. We start building a cluster around each isolated peak by selecting from detected centers. We decide whether or not to associate a center with a given cluster based on the collected statistics. Some heuristics may also be used to solve particular situations like the presence of a few contiguous centers that cannot be associated with any cluster.

In any case, whenever multiple hypotheses of clustering appear to be reasonable, all of them are taken into account, so that the *clustering subsystem* actually generates a set of hypotheses.

More precisely, let us assume that a cluster is represented by its *main peak*. This is defined as either the isolated peak contained in the cluster (if any) or any of the peaks of the cluster (possibly a central one).

Then, a *hypothesis of clustering* is the sequence of the main peaks of its component clusters:

$$\mathcal{H}^p = (x_{k_1^p}, x_{k_2^p}, \ldots, x_{k_{N_P}^p}) \tag{8}$$

where $x_{k_i^p}$ is the coordinate of the ith main peak in the pth hypothesis, while $N_P(p)$ is the number of clusters.

The *clustering subsystem* also sorts hypotheses according to their *clustering confidence*, which is expressed in terms of the global centering function values:

$$\mathcal{C}_C(\mathcal{H}^p) \triangleq \sum_{i=1}^{N_P(p)} \mathcal{F}(x_{k_i^p}) \tag{9}$$

5   Context Analysis Subsystem

For each hypothesis of clustering of the courtesy amount, starting from the one with the highest confidence, we produce a list of *hypothetical courtesy amounts*. These amounts are obtained by generating all the possible combinations of the pseudo-characters associated with each main peak, starting from combining the topmost alternatives in each list $\hat{\mathcal{L}}(x_{k_i^p})$.

The amounts are then sorted according to decreasing values of the *amount confidence*, which is calculated in terms of the weighted recognition confidences of the component pseudo-characters. Let $\mathcal{A}_m$ be the mth hypothetical courtesy amount. The amount confidence is

$$\mathcal{C}_A(\mathcal{A}^m, \mathcal{H}^p) \triangleq \frac{\sum_{i=1}^{N_P(p)} \hat{e}^{j(m,i)}(x_{k_i^p})}{N_P} \tag{10}$$

where $\hat{e}^j$ is the weighted recognition confidence, $j(m, i)$ indicates the index of the pseudo-character in the ith position of the mth amount, while p and m are the indexes of the hypothesis of clustering and the hypothetical courtesy amount, respectively.

Similar considerations also apply to legal amounts.

The hypothetical courtesy amount is used to divide the legal amount into smaller, easier to recognize, strings. Based on the number of digits in the courtesy amount, the presence of *characteristic tokens* can be assumed in the legal amount. For example, if the number of digits is greater than or equal to 4, the legal amount includes the token "mil" (for either "mille" or "mila").[1]

The characteristic tokens subdivide the legal string into substrings, which are, in turn, divided into lower-level substrings, and so on.

Actually, the system searches for *patterns*, i.e., sequences of pseudo-characters in the legal string which correspond to the searched tokens, after substitution of each character with the corresponding pseudo-character(s). As the translation of characters into pseudo-characters is not univocal, each characteristic token may have more than one equivalent pattern.

For each pattern, a *pattern belief* is computed in terms of the weighted recognition confidences associated with the pseudo-characters of that pattern. A *token belief* is then defined as the highest belief of the patterns associated with that token.

When all the tokens in the legal string have been detected, the check amount is identified. This amount is associated with an *amount confidence*, which is defined as the average of the token beliefs of its component tokens. Actually, as several hypotheses are considered, more than one check amount is identified. These amounts are sorted according to descending amount confidences.

6   Performance Evaluation

Table 2 lists the performance of the major processing blocks. It gives the RMS centering error for the CENTERING DETECTOR, the accuracy rates (with no rejection) for the CHARACTER RECOGNIZER and the *context analysis subsystem*. Accuracy rate can be improved if check rejection can be accepted (as is usually the case). Overall system performance

---

[1] Italian for "thousand/s"

Table 2. Performance of the HACRE system, measured on 80 checks.

Module	Parameter	Courtesy	Legal
CENTERING DETECTOR	RMS centering error (distance between detected center coordinate x_k and actual geometrical center of pseudo-character), in pixels.	2.5	3.0
CHARACTER RECOGNIZER	Average accuracy rate (correct, if the pseudo-character is in the first five positions in the list $\hat{\mathcal{L}}(x_k)$), with no rejection.	88%	65%
context analysis subsystem	Average accuracy rate (correct, if the correct amount is in the first 4 positions of the list of final amounts), with no rejection.	53%	

is evaluated as the percentage of amounts correctly recognized in the first k $(1 \leq k \leq 4)$ positions.

Table 3 lists execution times of the various processing blocks; figures are given for a system with 64 PEs running at 66 MHz and a Pentium with 90MHz clock frequency. It is clear that the system can recognize up to 3 checks per second.

Note that total times are much shorter than the sum of individual processing times, due to the pipelined internal architecture of PAPRICA-3 and the superposition of PAPRICA-3 and Pentium processing.

All PAPRICA-3 programs were also tested on a Pentium at 90 MHz, using the same algorithms based on mathematical morphology, which are well suited to the specific problems of bitmap processing and character recognition. Some programs (FILTER, BRIGHTNESS, THRESHOLD, ZOOM, CHARACTER RECOGNIZER) can be implemented more efficiently on a sequential computer using more traditional methods (ad-hoc programs). These were implemented on the Pentium and their performance is also listed in Table 3 for comparison.

Note that the performance of HACRE is two to three decades faster than that of Pentium, for almost all the programs considered.

Table 3. Average execution times of the various processing steps of HACRE, while processing the courtesy amount. [†] COMPRESS acts on an image zoomed by an average factor 3.2; therefore, processing times are scaled accordingly. [††] CHARACTER RECOGNIZER acts a few times per each pseudo-character, namely, once every 15 lines on average.

	PAPRICA-3, 64 PEs, 66 MHz		Pentium, 90 MHz	
	worst case		morphol.	ad-hoc
Image preprocessor	µs/line	ms/check	µs/line	µs/line
WINDOW EXTR.+FILTER	1.38	2.76	1,700	705
BRIGHTNESS	2.95	5.90	1,970	320
THRESHOLD	0.91	1.82	830	255
THINNING	8.34	16.7	7,390	←
BASELINE	4.24	8.28	4,320	←
FEATURES	3.05	6.10	9,490	←
ZOOM	2.24	4.48	820	160
COMPRESS[†]	30.8	61.6	21,350	←
OTHER (VARIOUS)	3.25	6.50	3,330	←
TOTAL PREPROCESSING	57.2	114.0	51,200	47,320

	PAPRICA-3, 64 PEs, 66 MHz		Pentium, 90 MHz	
	worst case		morphol.	ad-hoc
Neural subsystem	ms/psd-char	ms/check	ms/check	
CENTERING DETECTOR	0.40	19.2	1,440	-
RECOGNIZER (FEATURES)[††]	3.60	172.8	12,960	-
RECOGNIZER (NEURAL)[††]	1.68	80.7	-	13,440
TOTAL RECOGNIZER	5.68	272.7	27,840	

	Pentium, 90 MHz			
	ms/psd-char		ms/check	
	courtesy	legal	courtesy	legal
clustering subsystem	4.0	1.3	28.0	36.0
context analysis subsyst.	0.26		6.7	

PREPROC. + RECOGN. (PAPRICA-3) and *clustering + context anal.* (Pentium)	
ms/psd-char	ms/check
Total 7.15	343

7   Summary

This chapter has presented a high-speed handwriting recognizer for interpreting the amount on Italian bank checks (note that the system can easily be upgraded to deal with other languages). The system has an accuracy of 53% and can interpret up to 3 checks per second.

In the near future we foresee improvement of the performance of each individual block, and in particular, the *neural subsystem* and the *context analysis subsystem*. In addition, we foresee evaluation of system performance when accepting check rejection. We therefore expect a significant improvement in recognition accuracy, while maintaining the same processing speed.

References

[1] Cohen, E. (1994) Computational theory for interpreting handwritten text in constrained domains, *Artificial Intelligence*, Elsevier Science, New York, Vol. 67, pp. 1-31.

[2] Evett, E. et al. (1991) Using linguistic information to aid handwriting recognition, *Proc. Intl. Workshop on Frontiers in Handwriting Recognition*, pp. 303-311, September.

[3] Schenkel, M. et al. (1992) Recognition-based Segmentation of On-line Hand-printed Words, *Advances in Neural Information Processing System 3*, Morgan Kaufmann, pp. 723-730.

[4] Gregoretti, F. et al. (1996) The Implementation of the PAPRICA-3 Parallel Architecture, *Proc. of 2nd Intl. Conference on Massively Parallel Computing Systems*, IEEE Computer Society Press, pp. 87-94, May.

[5] Serra, J. (1992) Image Analysis and Mathematical Morphology, *Academic Press*, London.

[6] Bozinovic, R.M. and Srihari, S.N. (1989) Off-line cursive script word recognition, *IEEE Trans. on PAMI*, Vol. 11, pp. 68-83, January.

[7] Keeler J. and Rumelhart, D.E. (1992) A Self-organizing Integrated Segmentation and Recognition Neural Net, *Advances in Neural Information Processing Systems 4*, Morgan Kaufmann, pp. 496-503.

[8] Teulings, H.L. (1993) Invariant handwriting features useful in cursive-script recognition, *Fundamentals in Handwriting Recognition*, Impedovo, S., Ed., Springer-Verlag, pp. 179-198.

[9] Burges, C.J.C. et al. (1993) Off-line Recognition of Handwritten Postal Words Using Neural Networks, *Intl. Journal on Pattern Recognition and Artificial Intelligence*, Vol. 7, pp. 689-703.

[10] Reyneri, L.M. (1995) Weighted Radial Basis Functions for Improved Pattern Recognition and Signal Processing, *Neural Processing Letters*, May pp. 2-6.

[11] Haykin, S. (1994) *Neural Networks: A Comprehensive Foundation*, McMillan College Publishing Company, New York.

[12] Pirlo, G. (1993) Algorithms for Signature Verification, in *Fundamentals in Handwriting Recognition*, Impedovo, S., Ed., Springer-Verlag, Berlin, pp. 436-454.

Chapter 9:

Off-Line Handwritten Word Recognition Using Hidden Markov Models

OFF-LINE HANDWRITTEN WORD RECOGNITION USING HIDDEN MARKOV MODELS

A. El-Yacoubi,[1,4] **R. Sabourin,**[1,2] **M. Gilloux**[3] **and C.Y. Suen**[1]

[1] Centre for Pattern Recognition and Machine Intelligence
Department of Computer Science, Concordia University
1455 de Maisonneuve Boulevard West
Suite GM-606, Montréal, Canada H3G 1M8

[2] Ecole de Technologie Supérieure
Laboratoire d'Imagerie, de Vision et d'Intelligence Artificielle (LIVIA)
1100 Notre-Dame Ouest, Montréal, Canada H3C 1K3

[3] Service de Recherche Technique de La Poste
Département Reconnaissance, Modélisation Optimisation (RMO)
10, rue de l'île Mâbon, 44063 Nantes Cedex 02, France

[4] Departamento de Informatica (Computer Science Department)
Pontificia Universidade Catolica do Parana
Av. Imaculada Conceicao, 1155 - Prado Velho
80.215-901 Curitiba - PR - BRAZIL

This chapter describes a system that recognizes freely handwritten words off-line. Based on Hidden Markov models (HMMs), this system is designed in the context of a real application in which the vocabulary is large but dynamically limited. After preprocessing, a word image is segmented into letters or pseudo-letters and represented by two feature sequences of equal length, each consisting of an alternating sequence of shape-symbols and segmentation-symbols that are both explicitly modeled. The word model is made up of the concatenation of appropriate letter models which consist of elementary HMMs. Two mechanisms are considered to reject unreliable outputs, depending on whether or not the unknown word image is guaranteed to belong to the dynamic lexicon. Experiments performed on real data show that HMMs can be successfully used for handwriting recognition.

0-8493-9807-X/99/$0.00+$.50

1   Introduction

Today, handwriting recognition is one of the most challenging tasks and exciting areas of research in computer science. Indeed, despite the growing interest in this field, no satisfactory solution is available. The difficulties encountered are numerous and include the huge variability of handwriting such as inter-writer and intra-writer variabilities, writing environment (pen, sheet, support, etc.), the overlap between characters, and the ambiguity that makes many characters unidentifiable without referring to context.

Owing to these difficulties, many researchers have integrated the lexicon as a constraint to build lexicon-driven strategies to decrease the problem complexity. For small lexicons, as in bank-check processing, most approaches are global and consider a word as an indivisible entity [1] – [5]. If the lexicon is large, as in postal applications (city name or street name recognition) [6] – [10], one cannot consider a word as one entity, because of the huge number of models which must be trained.

Therefore, a segmentation of words into basic units, such as letters, is required. Given that this operation is difficult, the most successful approaches are segmentation-recognition methods in which a loose segmentation of words into letters or pieces of letters is first performed, and the optimal combination of these units to retrieve the entire letters (definitive segmentation) is then obtained in recognition using *dynamic programming* techniques [11], [12], [13]. These methods are less efficient when the segmentation fails to split some letters.

On the other hand, they have many advantages over global approaches. The first is that for a given learning set, it is more reliable to train a small set of units such as letters than whole words. Indeed, the frequency of each word is far lower than the frequency of each of its letters, which are shared by all the words of the lexicon. Furthermore, unlike analytic approaches, global approaches are possible only for lexicon-driven problems and do not satisfy the portability criterion, since for each new application, the set of words of the associated lexicon must be trained.

More recently, *hidden Markov models* (*HMMs*) [14], [15] have become the predominant approach to automatic speech recognition [16], [17], [18]. The main advantage of HMMs lies in their probabilistic nature, suitable for signals corrupted by noise such as speech or handwriting, and in their theoretical foundations, which are behind the existence of powerful algorithms to automatically and iteratively adjust the model parameters.

The success of HMMs in speech recognition has recently led many researchers to apply them to handwriting recognition, by representing each word image as a sequence of observations. According to the way this representation is carried out, two approaches can be distinguished: *implicit segmentation* [6], [19], [20], which leads to a speech-like representation of the handwritten word image, and *explicit segmentation* [7], [9] which requires a segmentation algorithm to split words into basic units such as letters.

In this chapter, we propose an explicit segmentation-based HMM approach to recognize unconstrained handwritten words (uppercase, cursive and mixed). An example of the kind of images to be processed is shown in Figure 1. This system uses *three* sets of features: the first two are related to the shape of the segmented units, while the features of the third set describe segmentation points between these units. The first set is based on global features such as loops, ascenders and descenders, while the second set is based on features obtained by an analysis of the bidimensional contour transition histogram of each segment. Finally, segmentation features correspond to either spaces, possibly occurring between letters or words, or to the vertical position of the segmentation points splitting connected letters. Given that the two sets of shape-features are separately extracted from the image, we represent each word by two feature sequences of equal length, each consisting of an alternating sequence of shape-symbols and segmentation-symbols.

In the problem we are dealing with, we consider a vocabulary which is large but dynamically limited. For example, in city name recognition, the contextual knowledge brought by the postal code identity can be used to reduce the lexicon of possible city names to a small size. However, since the entire lexicon is large, it is more realistic to model letters rather than whole words. Indeed, this technique needs only a reasonable number of models to train (and to store). Then each word (or word sequence) model can be built by concatenating letter models. This modeling is also more appropriate to available learning databases, which often do not contain all the possible words that need to be recognized.

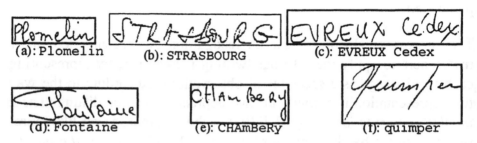

(a): Plomelin   (b): STRASBOURG   (c): EVREUX Cedex

(d): Fontaine   (e): CHAmBeRy   (f): quimper

Figure 1. Examples of handwritten word images of city names in France.

This chapter is organized as follows. Section 2 describes the theory of HMMs, and particularly emphasizes some variants that can enhance the standard modeling. Section 3 recalls the steps of preprocessing, segmentation and feature extraction. Section 4 deals with the application of HMMs to handwritten word recognition in a dynamic vocabulary. Section 5 presents the experiments performed to validate the approach. Section 6 concerns the rejection mechanism considered by our system. Finally, Section 7 presents some concluding remarks and perspectives.

2   Hidden Markov Models

During the last 15 years, HMMs have been extensively applied in several areas including speech recognition [18], [21], [22], [23], language modeling [24], handwriting recognition [6], [9], [25], [26], on-line signature verification [27], human action learning [28], fault detection in dynamic systems [29] and recognition of moving light displays [30].

A hidden Markov model is a doubly stochastic process, with an underlying stochastic process that is not observable (hence the word *hidden*), but can be observed through another stochastic process that produces the sequence of observations [14]. The hidden process consists of a set of *states* connected to each other by *transitions* with probabilities, while the observed process consists of a set of *outputs* or *observations*, each of which may be emitted by each state according to some output probability density function (*pdf*). Depending on the nature of this *pdf*, several kinds of HMMs can be distinguished. If the observations are naturally discrete or quantized using *quantization* or *vector quantization* [31], and drawn from an *alphabet* or a *codebook*, the HMM is called *discrete* [16], [17]. If these observations are continuous, we are dealing with a *continuous* HMM [17], [32], with a continuous *pdf* usually approximated by a mixture of normal distributions. Another family of HMMs, a compromise between discrete and continuous HMMs, are *semi-continuous* HMMs [33] that mutually optimize the vector quantized codebook and HMM parameters under a unified probabilistic framework.

In some applications, it is more convenient to produce observations by transitions rather than by states. Furthermore, it is sometimes useful to allow transitions with no output in order to model, for instance, the absence of an event in a given stochastic process. If we add the possibility of using more than one feature set to describe the observations, we must modify the classic formal definition of HMMs [17]. For discrete HMMs, we can do this by considering the following parameters:

T: length of the observation sequence O; $O = O_0, O_1, ..., O_{T-1}$, where $O_t = (O_t^0, O_t^1, ..., O_t^p, ..., O_t^{P-1})$, the observation O_t^p at time t being drawn from the pth finite feature set, and $p = 0, 1, ..., P - 1$.

N: number of states in the model.

M_p: number of possible observation symbols for the pth feature set.

$S = \{s_0, s_1, ... s_{N-1}\}$: set of possible states of the model.

$Q = \{q_t\}$, $t = 0, 1, ..., T - 1$; q_t: state of the process at time t.

$V_p = \{v_1^p, v_2^p, ..., v_M^p\}$ codebook or discrete set of possible observation symbols corresponding to the pth feature set.

$A = \{a_{ij}\}$, $a_{ij} = Pr(q_{t+1} = s_j \mid q_t = s_i)$: probability of going from state s_i at time t to state s_j at time $t + 1$, and at the same time producing a real observation O_t at time t.

$A' = \{a'_{ij}\}$, $a'_{ij} = Pr(q_t = s_j \mid q_t = s_i)$: probability of null transition from state s_i at time t to state s_j at time t, producing null observation symbol Φ. Note here that there is no increase over time since no real observation is produced.

$B_p = \{b_{ij}^P(k)\}$, $b_{ij}^P(k) = Pr(O_t^p = v_k^p \mid q_t = s_i, q_{t+1} = s_j)$: output *pdf* of observing the kth symbol in the pth feature set when a transition from state s_i at time t to state s_j at time $t + 1$ is taken. If we assume the P output *pdfs* are independent (multiple codebooks), we can compute the output probability $b_{ij}(k)$ as the product of the P output probabilities:

$$b_{ij}(k) = \prod_{p=0}^{P-1} b_{ij}^P(k) \tag{1}$$

where $\Pi = \{\pi_i\}$, $\pi_i = Pr(q_1 = s_i)$ is the initial state distribution. In general, it is more convenient to have predefined initial and final states s_0 and s_{N-1} that do not change over time. In this case, $\pi_0 = 1$ and $\pi_i = 0$ for $i = 1, 2, ..., N - 1$.

A, A' and B_p obey the stochastic constraints:

$$\sum_{j=0}^{N} [a_{ij} + a'_{ij}] = 1 \qquad \sum_{k=0}^{M_p-1} b_{ij}^P(k) = 1 \qquad p = 0, 1, ..., P-1 \tag{2}$$

Given a model, to be represented by the compact notation $\lambda = (A, A', B_p)$ where $p = 0, 1, ..., P - 1$, three problems must be solved.

1. Given an observation sequence $O = O_0, O_1, ..., O_{T-1}$ and a model λ, how do we compute the probability of O given λ, $Pr(O \mid \lambda)$? This is the *evaluation* problem.

2. Given the observation sequence $O = O_0, O_1, ..., O_{T-1}$ and the model λ, how do we find the optimal state sequence in λ that has generated O? This is the *decoding* problem.

3. Given a set of observation sequences and an initial model λ, how can we re-estimate the model parameters so as to increase the likelihood of generating this set of sequences? This is the *training* problem.

2.1   The Evaluation Problem

To compute $Pr(O \mid \lambda)$, we modify the well-known *forward-backward* procedure [17] to take into account the assumption that symbols are emitted along transitions, the possibility of null transitions, and the use of multiple codebooks. Hence, we define the *forward* probability $\alpha_t(i)$ as

$$\alpha_t(i) = Pr(O_0, O_1, ..., O_{t-1}, q_t = s_i \mid \lambda) \tag{3}$$

i.e., the probability of the partial observation sequence $O_0, O_1, ..., O_{t-1}$ (until time $t - 1$) and the state s_i reached at time t given the model λ. $\alpha_t(i)$ can be inductively computed as follows:

Initialization

$$\alpha_0(0) = 1.0$$
$$\alpha_0(j) = \sum_{i=0}^{N-1} a'_{ij} \, \alpha_0(i) \qquad j = 0, 1, ..., N-1 \tag{4}$$

given that s_0 is the only possible initial state.

Induction

$$\alpha_t(j) = \sum_{i=0}^{N-1} \left[a_{ij} \left(\prod_{p=0}^{P-1} b_{ij}^p (O_{t-1}) \right) \alpha_{t-1}(i) + a'_{ij} \, \alpha_t(i) \right] \tag{5}$$

$$j = 0, 1, ..., N-1 \qquad\qquad t = 1, ..., T$$

by summing over all states that may lead to state s_j, and picking the appropriate time of a transition depending on whether we are dealing with a real observation or a null observation.

Termination

$$Pr(O \mid \lambda) = \alpha_T(N-1) \tag{6}$$

given that s_{N-1} is the only possible terminal state. Similarly, we define the *backward* probability $\beta_t(i)$ by

$$\beta_t(i) = Pr(O_t, O_{t+1}, ..., O_{T-1} \mid q_t = s_i, \lambda) \tag{7}$$

i.e., the probability of the partial observation sequence from time t to the end, given state s_i was reached at time t and the model λ. $\beta_t(i)$ can also be inductively computed as follows:

Initialization

$$\beta_T(N-1) = 1.0$$

$$\beta_T(i) = \sum_{j=0}^{N-1} a'_{ij} \, \beta_T(j) \qquad i = 0, 1, ..., N-1 \tag{8}$$

given that s_{N-1} is the only possible terminal state.

Induction

$$\beta_t(i) = \sum_{j=0}^{N-1} \left[a_{ij} \left(\prod_{p=0}^{P-1} b_{ij}^p (O_t) \right) \beta_{t+1}(j) + a'_{ij} \, \beta_t(j) \right] \tag{9}$$

$$i = 0, 1, ..., N-1 \qquad\qquad t = 1, ..., T-1$$

Termination

$$Pr(O \mid \lambda) = \beta_0(0) \tag{10}$$

given that s_0 is the only possible initial state.

2.2   The Decoding Problem

The decoding problem is solved using a near-optimal procedure, the *Viterbi* algorithm [34], [35], by looking for the best state sequence $Q =$

$(q_0, q_1,..., q_T)$ for the given observation sequence $O = O_0, O_1, ..., O_{T-1}$. Again, we modify the classic algorithm [17] in the following way. Let

$$\delta_t(i) = \max_{q_0, q_1,..., q_{T-1}} Pr(q_0, q_1,..., q_t = s_i, O_0, O_1, ..., O_{t-1} | \lambda) \quad (11)$$

i.e., $\delta_t(i)$ is the probability of the best path that accounts for the first t observations and ends at state s_i at time t. We also define a function $\Psi_t(i)$, the goal of which is to recover the best state sequence by a procedure called *backtracking*. $\delta_t(i)$ and $\Psi_t(i)$ can be recursively computed in the following way:

Initialization

$$\delta_0(0) = 1.0$$

$$\Psi_0(0) = 0 \quad (12)$$

$$\delta_0(j) = \max_{0 \le i \le N-1} a'_{ij}\delta_0(i) \qquad j = 0, 1, ..., N-1$$

$$\Psi_0(j) = \text{argmax}_{0 \le i \le N-1} a'_{ij}\delta_0(i) \qquad j = 0, 1, ..., N-1$$

given that s_0 is the only possible initial state.

Recursion

$$\delta_t(j) = \max_{0 \le i \le N-1} \left[a_{ij} \left(\prod_{p=0}^{P-1} b_{ij}^p (O_{t-1}) \right) \delta_{t-1}(i); a'_{ij} \delta_t(i) \right] \quad (13)$$

$$\Psi_t(j) = \text{argmax}_{0 \le i \le N-1} \left[a_{ij} \left(\prod_{p=0}^{P-1} b_{ij}^p (O_{t-1}) \right) \delta_{t-1}(i); a'_{ij} \delta_t(i) \right] \quad (14)$$

$$j = 0, 1, ..., N-1 \qquad t = 1, 2, ..., T$$

Termination

$$P^* = \delta_T(N-1) \quad (15)$$

$$q_T^* = N-1 \quad (16)$$

given that s_{N-1} is the only possible terminal state.

Path recovering: Backtracking procedure

$$q_t^* = \Psi_{t+1}(q_{t+1}^*) \qquad t = T-1, T-2, \ldots, 0. \qquad (17)$$

As shown above, except for the backtracking procedure, Viterbi and forward (Equations (4) – (6)) procedures are similar. The only difference is that the summation is replaced by a maximization.

2.3   The Training Problem

The main strength of HMMs is the existence of a procedure called the *Baum-Welch* algorithm [16], [17] that iteratively and automatically adjusts HMM parameters given a training set of observation sequences. This algorithm, which is an implementation of the *EM* (*expectation-maximization*) algorithm [36] in the HMM case, guarantees the model to converge to a local maximum of the probability of observation of the training set according to the *maximum likelihood estimation* (*MLE*) criterion. This maximum depends strongly on the initial HMM parameters. To re-estimate HMM parameters, we first define $\xi_t^1(i,j)$, the probability of being in state s_i at time t and in state s_j at time $t + 1$, producing a real observation O_t given the model and the observation O, and $\xi_t^2(i,j)$, the probability of being in state s_i at time t and in state s_j at time t, producing the null observation Φ given the model and the observation O.

$$\xi_t^1(i,j) = Pr(q_t = s_i, q_{t+1} = s_j \mid O, \lambda) \qquad (18)$$

$$\xi_t^2(i,j) = Pr(q_t = s_i, q_t = s_j \mid O, \lambda) \qquad (19)$$

The development of these quantities leads to

$$\xi_t^1(i,j) = \frac{\alpha_t(i)a_{ij}\left(\prod_{p=0}^{P-1} b_{ij}^p(O_t)\right)\beta_{t+1}(j)}{Pr(O\mid\lambda)} \qquad (20)$$

$$\xi_t^2(i,j) = \frac{\alpha_t(i)a'_{ij}\beta_t(j)}{Pr(O\mid\lambda)} \qquad (21)$$

We also define $\gamma_t(i)$ as the probability of being in state s_i at time t, given the observation sequence and the model.

$$\gamma_t(i) = Pr(q_t = s_i \mid O, \lambda) \tag{22}$$

$\gamma_t(i)$ is related to $\xi_t^1(i,j)$ and $\xi_t^2(i,j)$ by the following equation:

$$\gamma_t(i) = \sum_{j=0}^{N-1} [\xi_t^1(i,j) + \xi_t^2(i,j)] = \frac{\alpha_t(i)\beta_t(i)}{Pr(O \mid \lambda)} \tag{23}$$

The re-estimations of HMM parameters $\{a_{ij}\}$, $\{a'_{ij}\}$, $\{b_{ij}^P(k)\}$ are

$$\overline{a_{ij}} = \frac{\text{expected number of transitions from } s_i \text{ at time } t \text{ to } s_j \text{ at time } t+1}{\text{expected number of being in } s_i} \tag{24}$$

$$\overline{a'_{ij}} = \frac{\text{expected number of transitions from } s_i \text{ to } s_j \text{ and observing } \Phi}{\text{expected number of being in } s_i} \tag{25}$$

$$\overline{b_{ij}^P(k)} = \frac{\text{exp. num. of symbols } v_k^P \text{ in trans. from } s_i \text{ at time } t \text{ to } s_j \text{ at time } t+1}{\text{exp. num. of transitions from } s_i \text{ at time } t \text{ to } s_j \text{ at time } t+1} \tag{26}$$

Given the definitions of $\xi_t^1(i,j)$, $\xi_t^2(i,j)$ and $\gamma_t(i)$, it is easy to see, when we are using one observation sequence O, that

$$\overline{a_{ij}} = \frac{\displaystyle\sum_{t=0}^{T} \xi_t^1(i,j)}{\displaystyle\sum_{t=0}^{T} \gamma_t(i)} = \frac{\displaystyle\sum_{t=0}^{T} \alpha_t(i) a_{ij} \left(\prod_{p=0}^{P-1} b_{ij}^P(O_t) \right) \beta_{t+1}(j)}{\displaystyle\sum_{t=0}^{T} \alpha_t(i)\beta_t(i)} \tag{27}$$

$$\overline{a'_{ij}} = \frac{\displaystyle\sum_{t=0}^{T} \xi_t^2(i,j)}{\displaystyle\sum_{t=0}^{T} \gamma_t(i)} = \frac{\displaystyle\sum_{t=0}^{T} \alpha_t(i) a'_{ij} \beta_t(j)}{\displaystyle\sum_{t=0}^{T} \alpha_t(i)\beta_t(i)} \tag{28}$$

$$\overline{b_{ij}^{p}(k)} = \frac{\sum_{t=0}^{T} \delta(O_{t}^{p}, v_{k}^{p}) \xi_{t}^{1}(i,j)}{\sum_{t=0}^{T} \xi_{t}^{1}(i,j)} = \frac{\sum_{t=0}^{T} \delta(O_{t}^{p}, v_{k}^{p}) \alpha_{t}(i) a_{ij} \left(\prod_{p=0}^{P-1} b_{ij}^{p}(O_{t}) \right) \beta_{t+1}(j)}{\sum_{t=0}^{T} \alpha_{t}(i) a_{ij} \left(\prod_{p=0}^{P-1} b_{ij}^{p}(O_{t}) \right) \beta_{t+1}(j)} \tag{29}$$

where $\quad \delta(x,y) = \left(\begin{array}{ll} 1 & \text{if } x = y \\ 0 & \text{otherwise} \end{array} \right)$

For a set of training sequences $O(0)$, $O(1)$,..., $O(U\text{-}1)$ (size U), as is usually the case in real-world applications, the above formulas become

$$\overline{a_{ij}} = \frac{\sum_{u=0}^{U-1} \sum_{t=0}^{T} \xi_{t}^{1}(i,j,u)}{\sum_{u=0}^{U-1} \sum_{t=0}^{T} \gamma_{t}(i,u)} = \frac{\sum_{u=0}^{U-1} \frac{1}{P(u)} \sum_{t=0}^{T} \alpha_{t}(i,u) a_{ij} \left(\prod_{p=0}^{P-1} b_{ij}^{p}(O_{t}^{p}(u)) \right) \beta_{t+1}(j,u)}{\sum_{u=0}^{U-1} \frac{1}{P(u)} \sum_{t=0}^{T} \alpha_{t}(i,u) \beta_{t}(i,u)} \tag{30}$$

$$\overline{a'_{ij}} = \frac{\sum_{u=0}^{U-1} \sum_{t=0}^{T} \xi_{t}^{2}(i,j,u)}{\sum_{u=0}^{U-1} \sum_{t=0}^{T} \gamma_{t}(i,u)} = \frac{\sum_{u=0}^{U-1} \frac{1}{P(u)} \sum_{t=0}^{T} \alpha_{t}(i,u) a'_{ij} \beta_{t}(j,u)}{\sum_{u=0}^{U-1} \frac{1}{P(u)} \sum_{t=0}^{T} \alpha_{t}(i,u) \beta_{t}(i,u)} \tag{31}$$

$$\overline{b_{ij}^{p}(k)} = \frac{\sum_{u=0}^{U-1} \sum_{t=0}^{T} \delta(O_{t}^{p}(u), v_{k}^{p}) \xi_{t}^{1}(i,j,u)}{\sum_{u=0}^{U-1} \sum_{t=0}^{T} \xi_{t}^{1}(i,j,u)} =$$

$$= \frac{\sum_{u=0}^{U-1} \frac{1}{P(u)} \sum_{t=0}^{T} \delta(O_{t}^{p}(u), v_{k}^{p}) \alpha_{t}(i,u) a_{ij} \left(\prod_{p=0}^{P-1} b_{ij}^{p}(O_{t}^{p}(u)) \right) \beta_{t+1}(j,u)}{\sum_{u=0}^{U-1} \frac{1}{P(u)} \sum_{t=0}^{T} \alpha_{t}(i,u) a_{ij} \left(\prod_{p=0}^{P-1} b_{ij}^{p}(O_{t}^{p}(u)) \right) \beta_{t+1}(j,u)} \tag{32}$$

In the above equations, the index u is introduced into α, β, ξ^{1}, ξ^{2} and γ to indicate the observation sequence $O(u)$ currently used. Note that a new quantity $P(u) = Pr(O(u)|\lambda)$ appears, since this term is now included in the summation and cannot be eliminated as before. Training can also be performed using the *segmental k-means* algorithm or Viterbi training [37]. The idea behind this algorithm is that after initializing the model parameters with random values, each word is matched against its associated feature sequence via the Viterbi algorithm. According to the current model, observation sequences of the training set are segmented into states (or transitions) by recovering the optimal alignment path. The re-estimations of the new HMM parameters are then directly obtained by examining the number of transitions between states and the number of observations emitted along transitions. This procedure is repeated (as in Baum-Welch training) until the increase in the global probability of observing training examples falls below a small fixed threshold. Although this algorithm is less optimal than the Baum-Welch algorithm, it generally leads to good results and is faster in computation.

3   Representation of Word Images

In Markovian modeling, each input word image must be represented as a sequence of observations, which should be statistically independent, once the underlying hidden state is known. To fulfill the latter requirement, the word image is first preprocessed by 4 modules: *baseline slant normalization, lower case letter area normalization* when dealing with cursive words, *character skew correction*, and finally, *smoothing*.

Indeed, beside the fact that these variabilities are not meaningful to recognition and cause a high writer-sensitivity in classification, thus increasing the complexity in a writer-independent handwriting recognizer, they can introduce dependence between observations. For instance, a word with a highly slanted baseline is likely to give rise to many segments (after the segmentation process) with incorrectly detected descenders. In the absence of the baseline slant, none of these descenders will be detected, hence the idea of dependence between observations when the writing baseline is not normalized. The same thought can be made about the character slant. After preprocessing, we perform segmentation and feature extraction processes to transform the input image into an ordered sequence of symbols (first assumption).

3.1   Preprocessing

In our system, the preprocessing consists of four steps: *baseline slant normalization*, *lower case letter area (upper-baseline) normalization* when dealing with cursive words, *character skew correction*, and finally, *smoothing*. The goal of the first two is to ensure a robust detection of ascenders and descenders in our *first* feature set. The third step is required since the *second* feature set shows a significant sensitivity to character slant (Section 3.3). Baseline slant normalization is performed by aligning the minima of the lower contour after having removed those corresponding to descenders and those generated by pen-down movements, using the *least square method* and some thresholding techniques. Upper-baseline normalization is similar, and consists of aligning the maxima of the upper contour after having filtered those corresponding to ascenders or uppercase letters. However, the transformation here is non-linear since it must keep the normalized lower-baseline horizontal. The ratio of the number of filtered maxima over the total number of maxima is used as an *a priori* selector of the writing style: either cursive or mixed if this ratio is above a given threshold (fixed at 0.4 after several trials) or uppercase, in which case no normalization is done. Character skew is estimated as the average slant of elementary segments obtained by sampling the contour of the word image, without taking into account the horizontal and pseudo-horizontal segments. Finally, we carry out a smoothing to eliminate noise appearing at the borders of the word image, and resulting from the application of the continuous transformations associated to the above preprocessing techniques in a discrete space (bitmap). Figure 2 shows an example of the steps of preprocessing. More details on the description of these techniques can be found in [38].

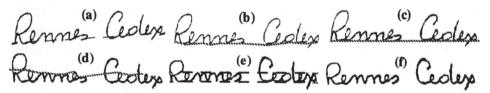

Figure 2. Preprocessing steps applied to word images: (a) original image, (b) and (c) baseline slant normalization, (d) character slant normalization, (e) lower case letter area normalization, (f) definitive image after smoothing.

3.2   Character Segmentation of Words

As mentioned earlier, segmentation of words into smaller units is necessary when dealing with large vocabularies. Segmentation techniques used in the framework of HMM-based handwritten word recognition approaches can be divided into *implicit* and *explicit* methods. Implicit methods are inspired by those considered in speech recognition which consist of sampling the speech signal into successive frames with a frequency sufficiently large to separately detect the different phonetic events (for instance, *phonemes*) using minimal supervised learning techniques [16], [39]. In handwriting, they can either work at the pixel column level [6], [20] or perform an *a priori* scanning of the image with sliding windows [19], [40]. Explicit methods, on the other hand, try to find explicitly the segmentation points in a word by using some characteristic points such as upper (or lower) contour minima, intersection points, or spaces. Implicit methods are better than explicit methods in splitting touching characters, for which it is hard to find regularly explicit segmentation points. However, due to the bidimensional nature of off-line handwritten word images, and to the overlap between letters, implicit methods are less efficient here than in speech recognition or on-line handwriting recognition. Indeed, vertical sampling makes it difficult to capture the sequential aspect of the strokes, which is better represented by explicit methods. Moreover, in implicit methods, segmentation points have to be also learned.

On the other hand, when employing explicit methods, the basic units to be segmented are naturally the alphabet letters. Unfortunately, because of the ambiguity encountered in handwritten words, it is impossible to correctly segment a word into characters without resorting to the recognition phase. Indeed, the same pixel representation may lead to several interpretations, in the absence of the context which can be a lexicon or grammatical constraints. In Figure 3, for instance, the group of letters inside the dashed square could be interpreted – in the absence of context given by the word Strasbourg – as "lreur," "lrun," "bour" (correct spelling), "baun," etc.

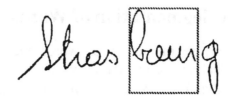

Figure 3. Ambiguity in handwritten words.

In view of the above remark, our concern is to design a segmentation process that tries to detect all the potential segmentation points, instead of only the real ones. This gives rise to several segmentation options, with the optimal one to be implicitly recovered during recognition. Integrating the above ideas, our segmentation algorithm is based on the following two hypotheses:

- There exist natural segmentation points corresponding to disconnected letters.

- The physical segmentation points between connected letters are located at the neighborhood of the image upper contour minima.

The segmentation algorithm makes use of upper and lower contours, loops, and upper contour minima. Then, to generate a segmentation point, a minimum must be located at the neighborhood of an upper-contour point that permits a vertical transition from the upper contour to the lower one without crossing any loop, while minimizing the vertical transition histogram of the word image. This strategy leads to a correct segmentation of a letter, to an undersegmentation of a letter (letter omission), or to an oversegmentation in which case a letter is split into more than one piece. Figure 4 gives an example of the behavior of the segmentation algorithm.

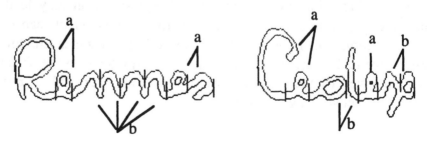

Figure 4. Segmentation of words into letters (a) or pseudo-letters (b).

3.3   Feature Extraction

The aim of the feature extraction phase is to extract in an ordered way, suitable to Markovian modeling, a set of relevant features that reduce redundancy in the word image, while preserving the discriminative information for recognition. Unlike in speech recognition where the commonly used features are obtained using well defined physical and mathematical concepts such as linear predictive coding (*LPC*) [17], there is no such agreement about the optimal features to be extracted from handwritten words. This is why topological features, features based on the pixel level, on the distribution of black pixels and on global transformations (*Fourier, Walsh, Karhunen-Loeve,* etc.) are often used in handwritten word recognition. Our main philosophy in this phase is that lexicon-driven word recognition approaches do not require features to be very discriminative at the segment (grapheme) level, because other information such as context (particular letter ordering in lexicon words, nature of the segmentation points) and word length, are available and permit high discrimination of words. Thus, our idea is to consider features at the grapheme level with the aim of clustering letters into classes.

Given our segmentation algorithm, a grapheme may consist of a full character, a piece of a character or more than a character. Such features cannot capture fine details of the segments, but this does allow, on the other hand, a description of the segments with less variability, ensuring a better learning of the distribution of the features over the characters.

In our system, the sequence of segments obtained by the segmentation process is transformed into a sequence of symbols by considering two sets of features. The first set [F_1 in Figure 6] is based on global features such as loops, ascenders and descenders. Ascenders (descenders) are encoded in two ways according to their relative size compared to the height of the upper (lower) writing zone. Loops are encoded in various ways according to their membership in each of the three writing zones (upper, lower, median), and their relative size compared to the sizes of these zones. The horizontal order of the median loop and the ascender (or descender) within a segment is also taken into account to ensure a better discrimination between letters such as "b" and "d" or "p" and "q."

This encoding scheme can be described simply by representing a segment by a binary vector, the components of which indicate the presence or the absence of the characteristics mentioned above. Each combination of these features within a segment is encoded by a distinct symbol. For example, in Figure 6, the first segment is encoded by symbol "L" reflecting the existence of a large ascender and a loop located above the core region. The second segment is encoded by symbol "o," indicating the presence of a small loop within the core region. The third segment is represented by symbol "–," which encodes shapes without any interesting feature. This scheme leads to an alphabet of 27 symbols.

The second feature set [F_2 in Figure 6] is based on the analysis of the bidimensional contour transition histogram of each segment in the horizontal and vertical directions. After a filtering phase, the histogram values may be 2, 4 or 6. We focus only on the median part of the histogram, which represents the stable area of the segment. In each direction, we determine the dominant transition number (2, 4 or 6). Each different pair of dominant transition numbers is then encoded by a different symbol or class. This coding leads to $3 \times 3 = 9$ symbols. In Figure 5, for instance, letters "B", "C" and "O", whose pairs of dominant transition numbers are (6,2), (4,2) and (4,4), are encoded by symbols called "B", "C" and "O", respectively.

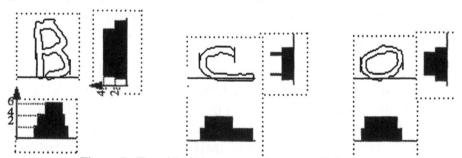

Figure 5. Transition histograms of segmented shapes.

In order to distinguish between the letters "A", "D", "O" and "P", ideally encoded by the same symbol "A" (4,4), we added new symbols by a finer analysis of the segments. The subclass ("O", "D") is chosen if the number of points of the lower contour located on the baseline and included in the median (or stable) zone of the segment is greater than a threshold, depending on the width of this zone. The subclass "P" is detected if the dominant number of transitions in the horizontal direction (4, in this case) loses its dominant character when focusing on the lower part of the segment. Similar techniques are used to discriminate between other letter classes, leading to a final feature set of 14 symbols.

In addition to the two feature sets describing segmented shapes, we also use segmentation features that try to reflect the way segments are linked together. These features consist of three categories. For not connected segments, two configurations are distinguished: if the space width is less than a third of the average segment width (*ASW*), we decide that there is no space and encode this configuration by the symbol "*n*"; otherwise, we validate the space and we encode it in two ways ("@" or "*#*"), depending on whether the space width is larger or smaller than the *ASW*.

For connected segments, the considered feature is the segmentation point vertical position. This feature is encoded in two ways (symbols "*s*" or "*u*") depending on whether the segmentation point is close to or far from the writing baseline. Hence, we obtain 5 segmentation features.

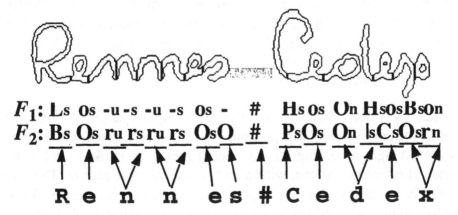

Figure 6. Pair of feature sequences representing a word (or sequence of words) image.

Space features have been considered in order to increase the discrimination between cursive and uppercase letters, while the vertical position of the segmentation or over-segmentation point is taken into account to discriminate between pairs of letters such as ("a","o"), ("u","v"), ("m","w"), ("H","U").

Finally, given that the two sets of shape-features are extracted independently, the feature extraction process represents each word image by two symbolic descriptions of equal length, each consisting of an alternating sequence of symbols encoding a segment shape and of symbols encoding the segmentation point associated with this shape (Figure 6).

4   Markovian Modeling of Handwritten Words

This section addresses the application of HMMs in handwritten word recognition. We begin first by briefly describing some related works in this field. Then, we give the justifications behind the design of the model we propose, and we detail the steps of learning and recognition as used in our system.

4.1   HMM Use in Handwritten Word Recognition

Recently, HMMs have been applied to several areas in handwriting recognition, including noisy printed text recognition [41], isolated character recognition [42], [43], on-line word recognition [44], [45], [46] and off-line word recognition. In the last application, several approaches have been proposed.

Gillies [6] is one of the first to propose an implicit segmentation-based HMM for cursive word recognition. First, a label is given to each pixel in the image according to its membership in strokes, holes and concavities located above, within and below the core region. Then, the image is transformed into a sequence of symbols which result from a vector quantization of each pixel column. Each letter is characterized using a different discrete HMM, the parameters of which are estimated on training data corresponding to hand-segmented letters. The Viterbi algorithm is used in recognition and allows an implicit segmentation of words into letters as a by-product of the word-matching process.

Bunke et al. [26] propose an HMM approach to recognize cursive words produced by cooperative writers. The features used in their scheme are based on the edges of the word skeleton graph. A semi-continuous HMM is considered for each character, with a number of states corresponding to the minimum number of edges expected of this character. The number of codebook symbols (the number of gaussians) was defined by manual inspection of the data, and recognition is performed using a *beam* search-driven Viterbi algorithm.

Chen et al. [47] use an explicit segmentation-based continuous density variable duration HMM for unconstrained handwritten word recognition. In this approach, observations are based on geometrical and topological features, pixel distributions, etc. Each letter is identified with a state which can account for up to 4 segments per letter. The statistics of the HMM (transition, observation and state duration probabilities) are estimated using the lexicon and the manually labeled training data. A modified Viterbi algorithm is applied to provide several outputs, which are post-processed using a general string editing method.

Cho et al. [40] use an implicit segmentation-based HMM to model cursive words. The word image is first split into a sequence of overlapping vertical grayscale bitmap frames, which are then encoded into discrete symbols using *principal component analysis* and vector quantization. A word is modeled by an interconnection network of character and ligature HMMs, with a number of states depending on the average sequence length of corresponding training samples. A clustering of ligature samples is performed to reduce the number of ligature models so as to ensure a reliable training. To improve the recognition strategy, several combinations of Forward and Backward Viterbi are investigated.

Finally, Mohamed and Gader [20] use an implicit segmentation-based continuous HMM for unconstrained handwritten word recognition. In their approach, observations are based on the location of black-white and white-black transitions on each image column. A 12-state left-to-right HMM is designed for each character. The training of the models is carried out on hand-segmented data, where the character boundaries are manually identified inside word images.

4.2   The Proposed Model

As shown above, several HMM architectures can be considered for handwritten word recognition. This stems from the fact that the correct HMM architecture is actually not known. The usual solution to overcome this problem is to first make structural assumptions, and then use parameter estimation to improve the probability of generating the training data by the models. In our case, the assumptions to be made are related to the behavior of the segmentation and feature extraction processes. As our segmentation process may produce a correct segmentation of a letter, a letter omission, or an oversegmentation of a letter into two or three segments, we built an eight-state HMM having three paths to take these configurations into account (Figure 7).

In this model, observations are emitted along transitions. Transition t_{07}, emitting the null symbol Φ, models the letter omission case. Transition t_{06} emits a symbol encoding a correctly segmented letter shape, while transition t_{67} emits a symbol encoding the nature of the segmentation point associated with this shape. Null transition t_{36} models the case of oversegmentation into only 2 segments. Transitions t_{01}, t_{23} and t_{56} are associated with the shapes of the first, second and third parts of an oversegmented letter, while t_{12} and t_{45} model the nature of the segmentation points that gave rise to this oversegmentation.

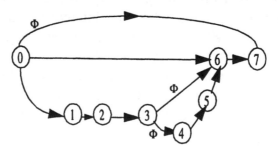

Figure 7. The character model.

This architecture is somewhat similar to that of other approaches such as [4], [47], but with some differences. In our method, the first segment presented to a character model is produced by two different transitions depending on whether it corresponds to the entire shape of a correctly segmented character (t_{06}) or to the first part of an oversegmented character (t_{01}), while in [4], [47] for example, the same transition is shared between these two configurations. The architecture proposed here allows the transitions of the model to be fed by homogeneous data sources, leading to less variability and higher accuracy (for example, the first part of an oversegmented "d" and a correctly segmented "d", which are very different, would be presented to different kinds of transitions (t_{01} and t_{06}, respectively). In other words, the variability coming from the inhomogeneity in the source data, since it is known *a priori*, is eliminated by separate modeling of the two data sources. We should also add that as each segment is represented in the feature extraction phase by two symbols related to our two feature sets, two symbols are independently emitted along transitions modeling segment shapes (t_{06}, t_{01}, t_{23} and t_{56}). In addition, we have a special model for inter-word space, in the case where the input image contains more than one word (Figure 1). This model simply consists of two states linked by two transitions, modeling a space (in which case only the symbols corresponding to spaces "@" or "#" can be emitted) or no space between a pair of words (Figure 8).

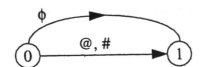

Figure 8. The inter-word space model.

4.3   The Learning Phase

The goal of the learning phase is to estimate the best parameter values of the character models, given a set of training examples and their associated word labels. Since the exact orthographic transcription (labeling) of each training word image is available, the word model is made up of the concatenation of the appropriate letter models; the final state of an HMM becomes the initial state of the next one, and so on (Figure 9).

The training is performed using the variant of the Baum-Welch procedure described in Section 2. Note here that we consider a *minimum supervised* training (given the exact transcription of words) in which the units (segments) produced by the segmentation algorithm need not be manually labeled by their associated letters or pseudo-letters. This is an important consideration for two reasons: first, labeling a database is a time-consuming and very expensive process, and is, therefore, not desirable; second, supervised training allows the recognizer to capture contextual effects, and permits segmentation of the sequence of units into letters and re-estimation of the transitions associated with these units to optimize the likelihood of the training database. Thus, the recognizer decides for itself what the optimal segmentation might be, rather than being heavily constrained by a priori knowledge based on human intervention [39]. This is particularly true if we bear in mind the inherent incorrect assumptions made about the HMM structure.

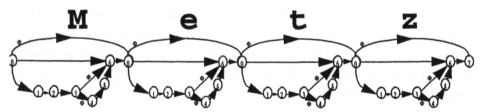

Figure 9. Training model for the French word Metz.

From an implementation point of view, given a word composed of L letters, a new parameter corresponding to the index of the currently processed letter is added to the quantities involved in the Baum-Welch algorithm. Then, the results of the final forward (initial backward) probabilities at the last (initial) state of the elementary HMM associated with a letter are moved forward (backward) to become the initial forward (final backward) probabilities at the initial (last) state of the elementary HMM associated with the following (previous) letter. If $\alpha_t^l(i)$ (or $\beta_t^l(i)$) denotes the standard forward (or backward) probability associated with the letter of index l, then this process is carried out according to the following equations:

$$\alpha_t^{l+1}(0) = \alpha_t^l(N-1) \qquad l = 0, ..., L-2 \qquad t = 0, 1, ..., T-1 \qquad (33)$$

$$\beta_t^{l-1}(N-1) = \beta_t^l(0) \qquad l = 0, ..., L-2 \qquad t = 0, 1, ..., T-1 \qquad (34)$$

s_0 and s_{N-1} being the initial and final states of elementary HMMs associated with letters. Similar modifications can be made to $\delta_t(i)$ if we want to use Viterbi training.

We should also add that the analysis of the segmentation process shows that splitting a character into three pieces is a rather rare phenomenon. Thus, the associated parameters are not likely to be reliably estimated, due to the lack of training examples that exhibit the desired events. The solution to this problem is to share the transitions involved in the modeling of this phenomenon (t_{34}, t_{36}, t_{45}, t_{56}) over all character models, by calling for the *tied states* principle. Two states are said to be tied when there exists an equivalence relationship between them (transitions leaving each of these two states are analogous and have equal probabilities) [16]. Nevertheless, this procedure is not carried out for letters M, W, m or w for which the probability of segmentation into 3 segments is high, and therefore, there are enough examples to train separately the parameters corresponding to the third segment for each of these letters. A further improvement consisted of considering *context-dependent* models for uppercase letters depending on their position in the word: first position whether in an uppercase or cursive word, or any different position in an uppercase word. The motivation behind this is that features extracted from these two categories of letters can be very different, since they are based on global features such as ascenders which strongly depend on the writing style by way of the writing baselines.

In addition to the learning set, we use a validation set in training on which the re-estimated model is tested after each iteration of the training algorithm. At the end of training, reached when the increase in the probability of generating the learning set by the models falls below a given threshold, the stored HMM parameters correspond to those obtained at the iteration, maximizing the likelihood of generating the validation set (and not the learning set) by the models. This strategy allows the models to acquire a better generalization over unknown samples.

4.4   The Recognition Phase

The task of the recognition problem is to find the word w (or word sequence) maximizing the *a posteriori* probability that w has generated an unknown observation sequence O:

$$Pr(\hat{w} \mid O) = \max_w Pr(w \mid O) \tag{35}$$

Applying Bayes' rule to this definition, we obtain the fundamental equation of pattern recognition,

$$Pr(w \mid O) = \frac{Pr(w \mid O)Pr(w)}{Pr(O)} \tag{36}$$

Since $Pr(O)$ does not depend on w, the decoding problem becomes equivalent to maximizing the joint probability,

$$Pr(w, O) = Pr(w \mid O)Pr(w) \tag{37}$$

$Pr(w)$ is the *a priori* probability of the word w and is directly related to the language of the considered task. For large vocabularies, $Pr(w)$ is very difficult to estimate due to the lack of sufficient training samples. The estimation of $Pr(O \mid w)$ requires a probabilistic model that accounts for the shape variations O of a handwritten word w. We assume that such a model consists of a global Markov model created by concatenating letter HMMs. The architecture of this model remains basically the same as in training. However, as no information in recognition is available on the style (orthographic transcription) in which an unknown word has been written, a letter model here actually consists of two models in parallel, associated with the upper and lower case modes of writing a letter (Figure 10). As a matter of fact, an initial state (I) and a final state (F) are considered, and two consecutive letter models are now linked by four transitions associated with the various ways two consecutive letters may be written: uppercase-uppercase (UU), uppercase-lowercase (UL), lowercase-uppercase (LU) and lowercase-lowercase (LL). The probabilities of these transitions are estimated by their occurrence frequency from the same learning database which served for HMM parameter estimation. The probabilities of beginning a word by an uppercase ($0U$) or lower-case letter ($0L$) are also estimated in the same way.

This architecture is more efficient than usually adopted methods which generate *a priori* two or three possible ASCII configurations of words (fully uppercase, fully lower-case or lower-case word beginning with an uppercase letter). Indeed, these methods quickly become tedious and time consuming when dealing with a word sequence rather than a single word, besides the fact that they cannot handle the problem of mixed handwritten words (e.g., Figure 1e). The proposed model elegantly avoids these problems, while the computation time increases only linearly. Recognition is performed using the variant of the Viterbi algorithm described in Section 2, allowing an implicit detection during recognition of the writing style which can be recovered by the backtracking procedure.

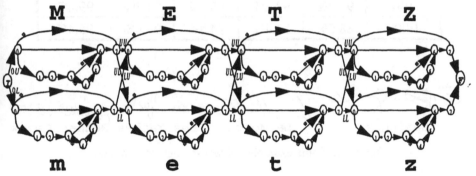

Figure 10. Global recognition model for lexicon word METZ.

5   Experiments

Experiments were carried out on unconstrained handwritten French city name images located manually on real mail envelopes. The sizes of the learning, validation and test databases were 11,106, 3,610 and 4,280, respectively. To simulate the address recognition task, we assume that the city names in competition are independent for a given list of corresponding postal code hypotheses. Under this assumption, for each image in the test set, we choose $N - 1$ city names from a vocabulary of 9,313 city names. The prior probabilities were assumed to be equal, so that the average perplexity was maximum and equal to N. In our tests, the values chosen for N were 10, 100 and 1,000. These values have a physical meaning since they simulate the case where 1, 2 or 3 digits in the postal code are ambiguous (1, 2 or 3 ambiguous digits give rise to 10, 100 or 1,000 possible postal codes). Recognition was carried out using the logarithmic version of the Viterbi procedure described in Section 2, while for training we used the Baum-Welch algorithm. Results of the tests are reported in Table 1, in which $RR(k)$ corresponds to the proportion of the correct answers among the k best solutions provided by the recognition module. Figure 11 shows some well-recognized images by our approach.

Table 1. Recognition rates obtained on various lexicon sizes.

Lexicon	RR(1)	RR(2)	RR(3)	RR(4)	RR(5)
10	99.0%	99.8%	99.9%	99.9%	100.0%
100	96.2%	98.2%	98.8%	99.2%	99.4%
1,000	88.5%	93.4%	94.8%	95.8%	96.4%

Epaignes   La Motte   Les Mathes Rennes Cedex

Figure 11. Some examples of well-recognized images.

The results shown in Table 1 prove that HMMs can be successfully used for designing a high-performance handwritten word recognition system. As mentioned in Section 3.3, even though the features used are not very discriminative at the grapheme level and do not capture fine details of the letters, the association (sequence) of these features to describe words in an ordered way is very discriminative, thanks to redundancy. This discrimination is as high as the length of the feature sequence extracted from the unknown word image. Furthermore, the description of the graphemes with features which do not take into account details, make these features more visible in the training set, thus ensuring a reliable estimation of the probability of observation of these features. Preprocessing and segmentation features also have been proven to significantly contribute to recognition accuracy [38]. It is not obvious to compare our approach with other works since we are not using the same databases, and also because our system recognizes actual French city names which may consist of one or several words. However, our results seem to compare favorably with others in the literature [40], [47]. Confusions in our system come mainly from poor images (Figure 12a), words with overlapping and touching characters (Figure 12b), words with truncated characters (Figure 12c), images with underline or with line above them (Figure 12d), or the lack of examples to reliably estimate some model parameters.

(a) Lorient   (b) Chartres   (c) Figeac   (d) Montmorillon

Figure 12. Some examples of misrecognized images.

6   Rejection

Usually, systems designed for real tasks are required to have *confusion rates* (*CR*) lower than some threshold depending on economical criteria. A typical value of accepted *CR* in postal applications is 1%, while a much lower value is required in the case of bank check processing. Therefore, it is necessary to consider in our approach a *rejection* criterion. In this perspective, we must go back to the Bayes formula in equation (36) to compute $Pr(O)$. When O is known to belong to the lexicon, as in our previous experiments, $Pr(O)$ can be obtained simply by

$$Pr(O) = \sum_{w} Pr(O \mid w) \times Pr(w) \tag{38}$$

Then rejection can be established by requiring $Pr(w|O)$ to be greater than a given threshold. Table 2 shows, when considering a dynamic lexicon of size 100, the evolution of the recognition rate and reliability (defined as the proportion of correct answers among the accepted images) as a function of the rejection rate by varying the threshold value, when the correct answer is guaranteed to belong to the lexicon.

Table 2. Recognition rate (*RC*) and reliability (*RL*) as a function of rejection rate (*RJ*) when the word image is guaranteed to belong to the lexicon.

RJ	0.0	1.9	4.3	7.0	7.7	8.6	9.0	9.8	10.9
RC	96.2	95.4	94.0	92.0	91.5	90.8	90.4	89.8	88.8
RL	96.2	97.2	98.2	98.9	99.1	99.3	99.4	99.5	99.7

In real applications, however, the processed word image is not guaranteed to belong to the lexicon, since it can be the result of a city name mislocation or a wrong dynamic generation of the lexicon (due to an important error in postal code recognition). To cope with this problem, we have randomly generated lexicons, *half* of which do not contain the correct answers. In this case, we express Bayes probability as proposed in [9] by

$$Pr(w|O) = \frac{p_{in} \times Pr(O|w) \times Pr(w)}{p_{in} \times \sum\limits_{w} Pr(O|w) \times Pr(w) + p_{out} \times Pr(O|out)} \tag{39}$$

where $p_{in} = 0.5$ is the *a priori* probability that w belongs to the lexicon and p_{out} is its complement; $p_{out} = 1 - p_{in}$. We approximated the term $Pr(O|out)$ by the output of an ergodic HMM trained using the Baum-Welch algorithm on the same set used to train the character models (although a more accurate set should also have included words that do not correspond to city names). The number of states of this HMM was set to 14 after several trials. Table 3 shows, for a dynamic lexicon of size 100, the evolution of the recognition rate and reliability as a function of the rejection rate, when the correct answer is not guaranteed to belong to the lexicon. Note that in this experiment, the recognition rate cannot exceed 50%, since $p_{in} = 0.5$.

Table 3. Recognition rate (*RC*) and reliability (*RL*) as a function of rejection rate (*RJ*) when the word image is not guaranteed to belong to the lexicon.

RJ	0.0	24.9	29.0	34.2	41.4	48.2	51.6	53.6	57.3
RC	48.0	47.5	47.1	46.7	45.8	44.0	42.8	42.0	39.6
RL	48.0	63.2	66.4	70.9	78.2	84.9	88.6	90.6	92.8

7 Summary

In this chapter, we described a complete system designed to recognize unconstrained handwritten words. The results obtained show that our approach achieves good performance, given that the data come from real-word images and that the writers were not aware that their words were to be processed by computer. One of the main strengths of our system lies in its training phase, which does not require any manual segmentation of the data. Due to the large size of the vocabulary, our Markovian modeling is carried out at the character level. Character

HMMs model not only segmented shapes, but also segmentation points, leading to better discrimination between letters. By building the word model as a sequence of character models, each consisting of a pair of associated uppercase and lower-case HMMs, the writing style is implicitly detected during recognition. An error analysis shows that our system can still be improved in most of its components. The segmentation algorithm should be optimized so as to be able to systematically split all the characters, particularly the overlapping and touching characters. Indeed, it is better to have pieces of characters which can be gathered during recognition, than segments containing more than one character. We also need more relevant features, since upper case letters, lower case letters, numerals and other special characters may be encountered in free-handwriting. This increases the level of ambiguity of the shapes generated by our segmentation algorithm, and therefore, a high level of description of these shapes is required. Moreover, given that our model assumes the independency between different feature sets (see Section 2), adding new feature sets does not increase the complexity from the training point of view and increases only linearly the required amount of memory to store the parameter values. A solution to avoid the inherent loss of information when generating our codebooks or feature sets is to replace discrete HMMs by semi-continuous HMMs. This can be particularly beneficial for our histogram-based features and segmentation features.

Acknowledgments

This work was supported by the Service de Recherche Technique de La Poste (SRTP) at Nantes, France, l'École de Technologie Supérieure (ETS) at Montréal, Canada, and the Centre of Pattern Recognition and Machine Intelligence (CENPARMI) at Montréal, Canada.

References

[1] Gilloux, M. and Leroux, M. (1992), "Recognition of cursive script amounts on postal cheques," *Proceedings of the 5th U.S. Postal Service Advanced Technology Conference*, pp. 545-556.

[2] Paquet, T. and Lecourtier, Y. (1993), "Recognition of Handwritten Sentences Using A Restricted Lexicon," *Pattern Recognition*, Vol. 26, No. 3, pp. 391-407.

[3] Gorski, N.D. (1994), "Experiments with Handwriting Recognition using Holographic Representation of Line Images," *Pattern Recognition Letters*, Vol. 15, No. 9, pp. 853-859.

[4] Knerr, S., Baret, O., Price, D., and Simon, J.C. (1996), "The A2iA Recognition System for Handwritten Checks," *Proceedings of Document Analysis Systems*, pp. 431-494.

[5] Suen, C.Y., Lam, L., Guillevic, D., Strathy, N.W., Cheriet, M., Said, J.N., and Fan, R. (1996), "Bank Check Processing System," *International Journal of Imaging Systems and Technology*, Vol. 7, pp. 392-403.

[6] Gillies, A.M. (1992), "Cursive Word Recognition Using Hidden Markov Models," *Proceedings of the 5th U.S. Postal Service Advanced Technology Conference*, pp. 557-562.

[7] Chen, M.Y., Kundu, A., and Zhou, J. (1994), "Off-Line Handwritten Word Recognition Using a Hidden Markov Model Type Stochastic Network," *IEEE Transactions on Pattern Analysis and Machine Intelligence*, Vol. 16, No. 5, pp. 481-496.

[8] Cohen, E., Hull, J.J., and Srihari, S.N. (1994), "Control Structure for Interpreting Handwritten Addresses," *IEEE Transactions on Pattern Analysis and Machine Intelligence*, Vol. 16, No. 10, pp. 1049-1055.

[9] Gilloux, M., Leroux, M., and Bertille, J.M. (1995), "Strategies for Cursive Script Recognition Using Hidden Markov Models," *Machine Vision and Applications*, Vol. 8, No. 4, pp. 197-205.

[10] El-Yacoubi, A., Bertille, J.M., and Gilloux, M. (1995), "Conjoined Location and Recognition of Street Names Within a Postal Address Delivery Line," *International Conference on Document Analysis and Recognition*, Vol. 2, pp. 1024-1027.

[11] Bozinovic, R.M. and Srihari, S.N. (1989), "Off-Line Cursive Word Recognition," *IEEE Transactions on Pattern Analysis and Machine Intelligence*, Vol. 11, No.1, pp. 68-83.

[12] Favata, J.T. and Srihari, S.N. (1992), "Cursive Word Recognition Using Hidden Markov Models," *Proceedings of the 5th U.S. Postal Service Advanced Technology Conference*, pp. 237-251.

[13] Kimura, F., Shridhar, M., and Chen, Z. (1993), "Improvements of a Lexicon Directed Algorithm for Recognition of Unconstrained Handwritten Words," *International Conference on Document Analysis and Recognition*, pp. 18-22.

[14] Rabiner, L.R. and Juang, B.H. (1986), "An Introduction to Hidden Markov Models," *IEEE Signal Processing Magazine*, Vol. 3, pp. 4-16.

[15] Poritz, A.B. (1988), "Hidden Markov Models: A Guided Tour," *IEEE International Conference on Acoustics, Speech, and Signal Processing*, pp. 7-13.

[16] Bahl, L., Jelinek, F., and Mercer, R. (1983), "A Maximum Likelihood Approach to Speech Recognition," *IEEE Transactions on Pattern Analysis and Machine Intelligence*, Vol. 5, No. 2, pp. 179-190.

[17] Rabiner, L.R. (1989), "A Tutorial on Hidden Markov Models and Selected Applications in Speech Recognition," *Proceedings of the IEEE*, Vol. 77, No. 2, pp. 257-286.

[18] Lee, K.F., Hon, H.W., Hwang, M.Y., and Huang, X. (1990), "Speech Recognition Using Hidden Markov Models: A CMU Perspective," *Speech Communication 9*, Elsevier Science Publishers B.V., North-Holland, pp. 497-508.

[19] Caesar, T., Gloger, J.M., Kaltenmeier, A., and Mandler, E. (1993), "Recognition of Handwritten Word Images by Statistical Methods," *Proceedings of the Third International Workshop on Frontiers in Handwriting Recognition*, pp.409-416.

[20] Mohamed, M. and Gader, P. (1996), "Handwritten Word Recognition Using Segmentation-Free Hidden Markov Modeling and Segmentation-Based Dynamic Programming Techniques," *IEEE Transactions on Pattern Analysis and Machine Intelligence*, Vol. 18, No. 5, pp. 548-554.

[21] Rabiner, L.R. (1989), "High Performance Connected Digit Recognition Using Markov Models," *IEEE Transactions on Acoustics, Speech and Signal Processing*, Vol. 37, No. 8, pp. 1214-1224.

[22] Averbuch, A., Bahl, L., Bakis, R., Brown, P., Daggett, G., Das, S., Davies, K., De Gennaro, S., De Souza, P.V., Epstein, E., Fraleigh, D., Jelinek, F., Lewis, B., Mercer, R., Moorhead, J., Nadas, A., Nahamoo, D., Picheny, M., Shichman, G., Spinelli, P., Van Compernolle, D. and Wilkens, H. (1987), "Experiments with the Tangora 20,000 word speech recognizer," *IEEE International Conf. on Acoustics, Signal and Speech Processing*, pp. 701-704.

[23] Chow, Y.L., Dunham, M.O., Kimball, O.A., Krasner, M.A., Kubala, G.F., Makhoul, J., Roucos, S., and Schwartz, R.M. (1987), "BIBLOS: The BBN Continuous Speech Recognition System," *IEEE International Conference on Acoustics, Speech, and Signal Processing*, pp. 89-92.

[24] Jelinek,F., Mercer, R.L., and Roukos, S. (1992), "Principles of Lexical Language Modeling for Speech Recognition," in Sadaoki Furui and M. Mohan Sondhi (Eds.), *Advances in Speech Signal Processing*, pp. 651-699.

[25] Kundu, A., He, Y., and Bahl, P. (1989), "Recognition of Handwritten Word: First and Second Order Hidden Markov Model Based Approach," *Pattern Recognition*, Vol. 22, No. 3, pp. 283-297.

[26] Bunke, H., Roth, M., and Schukat-Talamazzini, E.G. (1995), "Off-Line Cursive Handwriting Recognition using Hidden Markov Models," *Pattern Recognition*, Vol.28, No.9, pp. 1399-1413.

[27] Yang, L., Widjaja, B.K., and Prasad, R. (1995), "Application of Hidden Markov Models for Signature Verification," *Pattern Recognition*, Vol. 28, No. 2, pp. 161-170.

[28] Yang, J., Xu, Y., and Chen, S. (1997), "Human Action Learning via Hidden Markov Models," *IEEE Transactions on Systems, Man, and Cybernetics-Part A: Systems and Humans*, Vol. 27, No. 1, pp. 34-44.

[29] Smyth, P. (1994), "Hidden Markov Models for Fault Detection in Dynamic Systems," *Pattern Recognition*, Vol. 27, No. 1, pp. 149-164.

[30] Fielding, K.H. and Ruck, D.W. (1995), "Recognition of Moving Light Displays Using Hidden Markov Models," *Pattern Recognition*, Vol. 28, No. 9, pp. 1415-1421.

[31] Gray, R.M. and Lind, Y. (1982), "Vector Quantizers and Predictive Quantizers for Gauss-Markov Sources," *IEEE Transactions on Communications*, Vol. COM-30, No. 2, pp. 381-389.

[32] Liporace, L.A. (1982), "Maximum likelihood estimation for multivariate observation of Markov sources," *IEEE Transactions on Information Theory*, Vol. IT-28, No. 5, pp. 729-734.

[33] Huang, X.D. and Jack, M.A. (1989), "Semi-Continuous Hidden Markov Models for Speech Signals," *Computer Speech and Language*, Vol. 3, pp. 239-251, 1989.

[34] Forney, G.D. (1973), "The Viterbi Algorithm," *Proceedings of the IEEE*, Vol. 61, No. 3, pp. 268-278.

[35] Lou, H.L. (1995), "Implementing the Viterbi Algorithm," *IEEE Signal Processing Magazine*, pp. 42-52.

[36] Moon, T.K. (1996), "The Expectation-Maximization Algorithm," *IEEE Signal Processing magazine*, pp. 47-60.

[37] Rabiner, L.R. and Juang, B.H. (1993), *Fundamentals of Speech Recognition*, Englewood Cliffs, NJ: Prentice Hall, pp. 382-384.

[38] El-Yacoubi, A., Bertille, J.M., and Gilloux, M. (1994), "Towards a more effective handwritten word recognition system," *Proceedings of the Fourth International Workshop on Frontiers in Handwriting Recognition*, pp. 378-385.

[39] Picone, J. (1990), "Continuous Speech Recognition Using Hidden Markov Models," *IEEE Signal Processing Magazine*, pp. 26-41.

[40] Cho, W., Lee, S.W., and Kim, J.H. (1995), "Modeling and Recognition of Cursive Words with Hidden Markov Models," *Pattern Recognition*, Vol. 28, No. 12, pp.1941-1953.

[41] Elms, A.J. and Illingworth, J. (1998), "The Recognition of Noisy Polyfont Printed Text Using Combined HMMs," *International Journal on Document Analysis and Recognition (IJDAR)*, Vo. 1, No. 1, pp. 3-17.

[42] Nag, R., Wong, K.H., and Fallside, F. (1986), "Script Recognition Using Hidden Markov Models," *IEEE International Conference on Acoustics Signal and Speech Processing*, pp.2071-2074.

[43] Kim, H.J., Kim, K.H., Kim, S.K., and Lee, J.K. (1997), "On-Line Recognition of Handwritten Chinese Characters Based on Hidden Markov Models," *Pattern Recognition*, Vol. 30, No. 9, pp. 1489-1500.

[44] Bercu, S. and Lorette, G. (1993), "On-Line Handwritten Word Recognition: An Approach Based on Hidden Markov Models," *Proceedings of the Third International Workshop on Frontiers in Handwriting Recognition*, pp. 385-390.

[45] Ha, J.Y., Oh, S.C., and Kim, J.H. (1995), "Recognition of Unconstrained Handwritten English Words With Character and Ligature Modeling," *International Journal on Pattern Recognition and Artificial Intelligence*, Vol. 9, No. 3, pp. 535-556.

[46] Hu, J., Brown, M.K., and Turin, W. (1996), "HMM Based On-Line Handwriting Recognition," *IEEE Transactions on Pattern Analysis and Machine Intelligence*, Vol. 18, No.10, pp. 1039-1045.

[47] Chen, M.Y., Kundu, A., and Shrihari, S.N. (1995), "Variable Duration Hidden Markov Model and Morphological Segmentation for Handwritten Word Recognition," *IEEE Transactions on Image Processing*, Vol. 4, No. 12, pp. 1675-1687.

[48] El-Yacoubi, A., Sabourin, R., Gilloux, M. and Suen, C.Y. (1998), "Improved Model Architecture and Training Phase in an Off-line HMM-based Word Recognition System," *Proceedings of the 14th International Conference on Pattern Recognition*, pp. 1521-1525.

Chapter 10:

Off-Line Handwriting Recognition with Context-Dependent Fuzzy Rules

OFF-LINE HANDWRITING RECOGNITION WITH CONTEXT-DEPENDENT FUZZY RULES

A. Malaviya
Yatra Corporation
Suite 110, 5151 Edina Ind. Blvd.
Minneapolis, MN55439, U.S.A.
e-mail: Malaviya@computer.org

F. Ivancic, J. Balasubramaniam and **L. Peters**
Institute for System Design Technology
GMD – German National Research Center for Information Technology
D-53754 Sankt Augustin, Germany

The problem of off-line word recognition involves uncertainties at various levels. The fuzzy logic methodology is employed for handwriting analysis and recognition according to various uncertainties, from low-level preprocessing and segmentation tasks to word interpretation according to context. In this chapter, a multi-level fuzzy word recognition methodology is introduced. In the learning process, isolated or segmented handwriting portions are identified and stored in a form of knowledge base. This knowledge base is a combination of a rule base in a fuzzy pattern description language and the corresponding contextual information. A rule base in this language is automatically generated with the help of handwriting samples available for learning.

1 Methodology and Summary of Previous Work

A globally applicable off-line handwriting recognition system should recognize the handwriting of people with various backgrounds, nationalities, sex, professions and education, with equal reliability and competence. Most problems of current handwriting recognition systems lie in their extreme sensitivity to slight variability of handwriting styles. Lack of the ability to be reconfigured with changing environments, new characters, words and writers, is the other major problem. In the methodology presented here, ambiguities and variations of handwritten texts are processed with the help of fuzzy logic.

The first phase of the handwriting recognition process is learning, in which a knowledge base is automatically generated with sample handwriting data [3]. The knowledge base consists of a fuzzy rule base and corresponding contextual information in the form of a dictionary. A fuzzy combination of character and segment level information is aggregated here with the help of syntactic and semantic information. The syntactic information is obtained at the character level. Here, a list of possible characters and unidentified segments with important tagged features is generated. This is done with the help of extracted features and rules. In the second phase, the unidentified word images are classified with this knowledge base [5].

In this chapter, a methodology that divides the problem of handwriting recognition into various levels is described. The basic idea of multi-level fuzzy rule based pattern recognition is discussed in [8]. For word recognition, we extended the previous work of character recognition and added contextual information for semantic interpretation of words.

The human visual sense is very complex. It is selectively activated to various situations, and adapts quickly to unacquainted scenarios with some learning and compromises. The important properties of human visual systems are to adapt to local distortions in features, and more important to ignore unwanted details or redundant information [ORPE94]. This biological fact has motivated us to imitate such intelligence using multilevel fuzzy rules in this handwriting recognition system.

In the proposed method of handwriting recognition, tasks are divided into many levels according to the varying nature of uncertainties (see Figure 1). Each of these levels represents a linguistic space connected to other levels with fuzzy connectives [8]. The acquired handwriting data first undergoes segmentation, which also includes a preprocessing step. Picture uncertainties like image distortion, noise and other complex problems are handled using conventional algorithms and some fuzzy 'if-then' rules [18]. Feature extraction is the next level. Here, the segmented portions are identified or labeled for the appropriate features, with a corresponding membership function. In the aggregation phase, various types of features involving position, geometry and structure are combined to provide more meaningful features [10]. These new features cover various structural uncertainties including irregular shapes [11].

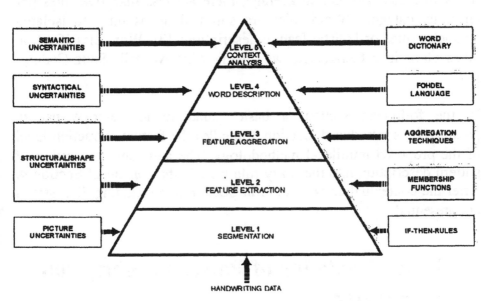

Figure 1. Fuzzy Word Recognition: a conceptual overview.

The next level is based on syntactical uncertainties, which is the relationship of identified features with each other. This relationship may be unequivocal. In this case, it has to be described in a way that variations in handwriting can be used for pattern description as well as for recognition. The interpretation of semantic uncertainties is highly context dependent and is part of the knowledge base with pattern description from the previous level. Later, unknown words are classified as possible natural language words, with corresponding membership values. This is done using the trained knowledge base. It is also possible to recognize parts of words and symbols using this method [13] [14].

The proposed method has been applied for sorting addressed letters for postal applications, i.e., ZIP Code, city name and street name recognition. The names of cities, ZIP codes and street names are obtained from special dictionaries. The important module of the proposed method is the extraction of meaningful features that characteristically represent the word image. The other important aspect is the automatic description of the unknown patterns in the form of a fuzzy rule base [9]. On an average, there are less than five rules per character category. Word rules are summed up as links to isolated character rules or holistic feature descriptions [3]. Word classification takes place in a hierarchical way using what we call 'Spitz-Coding' [17].

In the following section, a basic paradigm to describe pattern parameters is introduced. Section 3 handles the feature extraction level of the proposed multilevel methodology. The automatic generation of pattern description into the fuzzy rule base is the content of Section 4. After the contextual classification in Section 5, concluding discussions are presented.

2   Fuzzy Modeling of Pattern Description Parameters

Objects encountered in the real world generally do not have precisely defined criteria of membership [19]. For example, in the case of geometrical objects in an image pattern, a roughly described straight line may not clearly belong to the straight line class (which contains horizontal lines, vertical lines, positive or negative slanted lines) [7]. It may have some grade of membership in one or all of these classes depending on its inclination or orientation.

Fuzzy logic deals with a so-called fuzzy set F of the universe of discourse X, where the transition between full membership and no membership is gradual rather than abrupt. It may be seen as an extension of the range of the characteristic function from the discrete set $\{0,1\}$ to the continuous set $[0,1]$ [6]. Fuzzy set theory is, therefore, a generalization of conventional set theory [4].

Definition: Let X be a collection of objects x, which could be discrete
$$F=\{(x,m_F(x))\,|\,x\in X\},$$
or continuous, then a *fuzzy set* F in X is represented as a set of pairs where $m_F(x)$ denotes the membership function and maps an element x of the universe of discourse X to a real-numbered value in the interval $[0,1]$.

An essential role of the fuzzy modeling technique is the idea of a linguistic variable. In the linguistic approach, words or sentences are used in place of numbers to describe phenomena that are too complex or ill defined to be characterized in quantitative terms. A role comparable to that of a unit of measurement is played by one primary fuzzy set from which other sets can be generated by using linguistic modifiers such as "very", "quite", "more or less", etc. A property such as circularity can be measured in terms of "very circular", "more or less circular" or "a perfect circle". A linguistic variable can be either regarded as a variable whose value is a fuzzy number or as a variable whose values are defined by linguistic terms.

Definition: Let T be an injective function that maps an element t out of a set of names Σ into the set of all fuzzy membership functions of the universe of discourse X
$$M_F(X)=\{m_F\,|\,m_F:X\to[0,1]\}.$$
We call $t\in\Sigma$ a *linguistic term* and T(t) the corresponding membership function.

Fuzzy rule-based systems for pattern recognition and image understanding use a linguistic approach to overcome the information uncertainties. Handwritten symbols that have a certain deviation from the prototype characters can be a challenge to conventional recognition methods.

Methods for describing patterns in a linguistic form have been presented by various researchers over the last 30 years [7], [16], [20]. A complex pattern can often be described in terms of basic primitives and subpatterns. But the precision of formal languages in pattern recognition contradicts the imprecision or ambiguity of real life patterns. To overcome this difficulty, it is natural to introduce an uncertainty factor or fuzziness into the structure of formal languages [2]. This leads to the development of stochastic and fuzzy languages.

Fuzzy grammars are employed to describe the syntax of languages, and these grammars can be used to represent the structural relations of patterns [7], [18]. From the multitude of possible grammars, an attributed fuzzy grammar is chosen. This is due to its additional uncertainty description power over the conventional grammars.

Definition: A *fuzzy grammar* G_F is a quintuple

$$G_F = (V_N, V_T, P, S_0, m),$$

where

- V_N and V_T are finite disjoint sets of nonterminal and terminal vocabulary correspondingly, such that $V = V_N \cup V_T$ is the total vocabulary of the grammar.
- P is a finite set of production rules of the type $\alpha \rightarrow \beta$, where $\alpha \in V_N$ and β is member of the set V^* of all strings.
- $S_0 \in V_N$ is the starting symbol.
- We denote by m a mapping of P into [0,1], such that m(p) denotes the weighting factor of the rule $p \in P$.

We define a set P^* and a weighting function $m_P: P^* \rightarrow [0,1]$ for a unique fuzzy grammar recursively, as follows:

- For $\alpha \in V^*$, we have $\alpha \rightarrow^0 \alpha \in P^0$ and $m_P(\alpha \rightarrow^0 \alpha) = 1$.
- For $\alpha, \delta \in V^*$, $n \in \aleph$, we have

$$\alpha \rightarrow^{n+1} \delta \in P^{n+1} :\Leftrightarrow \exists A, C, \gamma \in V^*, \beta \in V_N :$$

$$\alpha \rightarrow^n A\beta C \in P^n \wedge \beta \rightarrow \gamma \in P \wedge A\gamma C = \delta.$$

In that case, the following is valid:

$$m_P(\alpha \to^{n+1} \delta) = m_P(\alpha \to^n A\beta C) \cdot m(\beta \to \gamma).$$

- We define the sets P^* by $P^* = U_{n \in \aleph} P^n$ and with $\alpha, \beta \in V^*$:
  $\alpha \to^* \beta :\Leftrightarrow \exists n \in \aleph : \alpha \to^n \beta \in P^n$.

A *fuzzy language* $L(G_F)$ is constructed with a unique fuzzy grammar as

$$L(G_F) = \left\{ (x, m_L(x)) \mid x \in V_T^*, S_0 \to^* x \right\}$$

where $m_L(x) = m_P(S_0 \to^* x)$. Therefore, m_L is a function that maps a terminal string x to a real-numbered value out of the interval [0,1].

To support the symbol classification process by means of fuzzy set theory, a fuzzy description language was developed. FOHDEL[1] incorporates fuzzy feature primitives, fuzzy linguistic terms and modifiers and fuzzy operators in the grammatical inference process [7] [13]. It provides a framework for describing handwritten symbols for recognizing and storing handwriting information.

3 Feature Extraction and Aggregation

Shape is considered as the primal geometrical property for extracting pattern recognition features. Daily life objects are recognized mostly due to their specific shapes. The perception of shapes can be viewed as a collective cognition of properties like size, form, symmetry and orientation. The possibilities to represent the geometrical shapes in mathematical terms have been formally investigated in [15]. In this study, it was considered that geometrical shapes in real life patterns have inherent uncertainties. Due to this, they cannot be accurately defined using conventional geometrical methods and crisp parameters. Thus a fuzzy approach to shape analysis, incorporating imprecise concepts, merits consideration.

The definition of a shape in terms of linguistic description like "curve is long/ round/ thin, etc." is easily related to known patterns by humans [11]. However, it is quite vague to define the shape of objects by these

---

[1] Fuzzy On-Line Handwriting Description Language

descriptors, as there is an obvious incompatibility with numerical processing methods. A closer look at these descriptors shows that geometrical, positional and global features furnish some insight about the shape itself, provided that they can be related to linguistic attributes or terms. For example, "curve", "circle" or "line" have a well defined geometric description. However, definitions of terms like "long", "thin" and "round" are more difficult to express by formulas without the semantic power offered by fuzzy set theory. Thus if we look at the circle example again, the description "the circle is round" is a fuzzy syntactic rule. Here, "circle" is the linguistic variable and "round" is one of the possible linguistic terms associated to the circle with a membership function. Our proposed approach collects the shape information through the extraction of geometric, positional and global features, and expresses this information in a set theoretic fuzzy manner.

In the segmentation phase, the input picture is binarized, filtered, transformed for orientations and the baseline is detected. In the actual segmentation phase, sharp corners, ends and crossings are considered to be nodes of segments. Global features characterize the word as a whole (like in the Spitz-Code) and geometric and positional features describe local aspects for each identified segment [18]. A detailed analysis of feature generation algorithms and aggregated features can be found in [11] and [7]. What follows is a brief glimpse of features utilized in the current work.

3.1 Positional Features

Positional features determine the relative position of the global or geometrical feature in the given rectangular window and are defined in the universe of discourse [0,1]. The two-dimensional universe of discourse is divided into six linguistic terms: {Left, Center, Right} and {Top, Middle, Bottom} (see Figure 2). These are associated to the fuzzy linguistic variables 'Vertical Position' (VP) and 'Horizontal Position' (HP), and express the relative position of a point, region, or segment to the centroid (xm,ym) of the analyzed character. These linguistic terms can be combined to create additional linguistic terms like "circle to the left-center" and "line right-top-center", which could be used to describe a "d". By just interchanging the term "left" with "right", the terms "circle to the right-center" and "line left-top-center" would be created to describe a "b". The global features describe the character as a whole and can be used for its preclassification. These shape features represent the global approximation of the shape. In other words, shape information from all portions combine to form the global description.

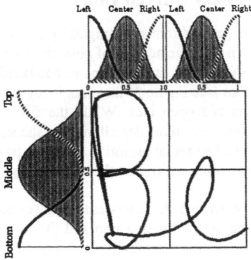

Figure 2. Positional features.

3.2 Discrimination of Straight Lines and Curved Lines

The shape of a given segmented subpattern can be classified to belong to the classes "straight line" or "arced line". The measure of 'straightness' of a segment is determined by fitting a straight line with the minimum least square error. Similarly, in a given segment, the ratio of the distance between end points and its total arc length shows its 'arced-ness'. In other words, if the distance between the end points is nearly equal to the total arc length, it is possibly a straight line. If the arc length is much greater than the distance between the end-points, it is very likely a curve. Here it is assumed that the curve is monotonous.

Once we have estimated the possibility of a segment to belong to a straight-line class, we have to identify the straight-line feature to which it belongs. If a segment is identified as a straight line, then the orientation or the angle of inclination of this segment classifies it further into one of the following features: vertical line, horizontal line, positive slant and negative slant.

3.3 Class of Fuzzy Curved Lines

Of the possible shapes and forms, distinguishing curved lines is a more complex task. From the large number of possible curved shape features, we selected the shapes that are frequent in handwriting patterns. We divided these curves (see Figure 3) into four categories: circles, S or Z type curves, loops and open arcs. While the first two classes each correspond to a linguistic term describing the shape, loops and open circles are described by several primitive linguistic terms depending on their crossing point or the position of their starting and ending points.

Most curves in handwritten characters are wholly or partially monotone convex. Depending on the missing part and the length of the arc, we can define them as: 'vertical curves' ($\supset,\subset$), 'horizontal curves' ($\cup,\cap$), 'hockey sticks' and 'walking sticks' (see Figure 3). The distinction of these shape categories is accomplished by using the angle of rotation, the angle of slope joining the end points of the segment, measure of arced-ness, the relative length and the area covered by the segment. All of these measures are relative and normalized in the fuzzy domain of discourse [0,1], and can be combined with the fuzzy aggregation operators. The output of this evaluation can classify a segment into several features and by this, associate several possible meanings to its shape. The decision that it belongs to one of these classes can be made at a later stage by adding contextual information or linking it to global or segment related features. In case of unavailability of a contextual relation, the class with the maximum grade of membership is considered the best choice. In what follows we present the structure of the feature primitives and the corresponding fuzzy membership functions. The categories 'horizontal curves' ($\cup,\cap$) and 'vertical curves' ($\supset,\subset$) can be described by convexity. In a vertical curve, a vertical line joins the end points of the curve and in a horizontal curve, a horizontal line joins the end points.

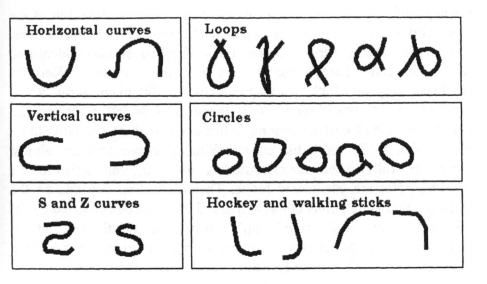

Figure 3. Curved features.

3.4 Combination of Features with Fuzzy Aggregation

The main objective of the fuzzy aggregation mechanism is to find an overall measure for certain fuzzy information from uncertain and imprecise information data. For selecting the meaningful features from a given set of handwriting data, it is necessary to consider the possible relationships among these features. The form of the combination is described with various fuzzy connectives, as explained in [7]. At segment level, various feature types like positional, line and curve categories are combined with each other. At character level the aggregated features are compared and features that are relatively more characteristic are selected.

From the simple monotone convex curves we can calculate aggregated features like hockey or walking sticks. These curves form the links between the horizontal and vertical feature curves in our universe of discourse. The curves are thus estimated through the aggregation of these features. The discriminatory component taken into consideration is the angle of the straight line connecting the end points. Like in the class of straight lines, this slant can be positive or negative. If the angle of the line is the main decision criterion for the class of straight lines, the angle of the line connecting the end points and the direction of the convexity is the corresponding parameter for the class of curved lines. Circles are a special form of curved lines. They can be easily identified with the measure of arced-ness as "very very high arced-ness". Here the attribute 'very-very-high' corresponds to the linguistic term VVH. Similarly, definitions of circles with openings to the left, right, top and bottom are extended by combining O-like curve with its similarity to C-like, D-like, A-like and U-like curves. While all curves that are "almost closed" or "fully closed" are classified as circles, the curves that have a crossing point are considered loops. Their definition depends on the starting and ending points of the loops, and they can be classified by the linguistic terms: 'left', 'right', 'up', 'down', 'loop-to-the-left', etc. S and Z curves are special curves, as they are neither concave nor convex, though their specific feature is their starting and ending points.

With this extraction and aggregation process, we have reached the level of syntactic uncertainty where structural information is in the form of features (see Figure 4). At this level, we have to understand these features in combination with each other. For this, we have to first generate a rule base that maps relational information from the extracted features. This is described in the following section. The generated rule base is then used to recognize characters and words. The automatic pattern description paradigm is a generic method for generating rules for characters as well as for words.

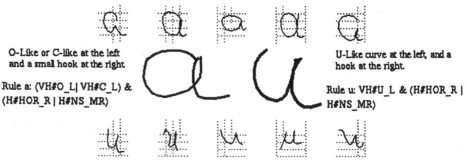

O-Like or C-like at the left and a small hook at the right.

Rule a: (VH#O_L| VH#C_L) & (H#HOR_R | H#NS_MR)

U-Like curve at the left, and a hook at the right.

Rule u: VH#U_L & (H#HOR_R | H#NS_MR)

Figure 4. Discrimination of characters with feature description.

4   Automatic Pattern Description Paradigm

The general idea of our approach to automate the generation of a fuzzy rule base is to begin by using established fuzzy clustering algorithms and to then enhance them. The generation process is initiated by using given data samples. Due to the high dimension of the input data, multi-phased clustering is introduced [3]. The idea of this new, multi-phased clustering approach is to learn the individual properties of each pattern class instead of the traditional way of learning the differences between the pattern classes. The rule base generation method is divided into two parts. The first part generates the initial set of rules for each input pattern class. Each rule consists of some specific features describing the pattern class. Fuzzy linguistic terms are associated to these features to describe their quality to discriminate the given pattern class. The second part of the algorithm cross checks the generated rules to find possible overlaps. The overlapping rules are altered until their classification output is separable or a certain iteration threshold is reached. In the second case, the overlapping rules are listed. The output of the proposed algorithm is a rule base written in FOHDEL [3]. This algorithm is summarized in Figure 5.

> **Input:** • set of patterns A that have to be classified
> • for each pattern fuzzy feature data in p
>   dimensions
> **Output:** fuzzy rule base in FOHDEL
> **Method:**
>   for each pattern do
> (1)   Sample Grouping
>   for each group do
> (2)   Reduction of Clustering Features
> (3)   Multiphased Clustering
> (4)   Rule Generation in FOHDEL
> (5)   od
>   od
> (6) Rule Cross-Checking and Refinement

Figure 5. The automatic rule base generation algorithm.

4.1 Multiphased Clustering

The first part of the algorithm is called multiphased clustering, because multiphased clustering is the most essential step in this part. The algorithm generates an initial set of rules for each pattern class from an unknown number n of samples of one class. Each sample corresponds to a feature vector, which contains the specific membership values of the sample for all predefined discriminatory features. These can be global or local in nature. The rule generation process is independently accomplished for each class.

The number of classes and features describing each class can be very large. This means that the number of clusters needed will also be very high. To have the rule generation process converge in a given time and to reduce the complexity of the problem, several data reduction steps are introduced in the proposed algorithm. Let the overall set of pattern data Γ be

$$\Gamma = \left\{ X^{(\gamma_1)}, X^{(\gamma_2)}, \ldots, X^{(\gamma_a)} \right\}$$

where $A = \{\gamma_1, \gamma_2, \ldots, \gamma_a\}$ is the set of patterns, that have to be recognized, with $\forall\ i,j \in \{1, \ldots, a\}: i \neq j \Rightarrow \gamma_i \neq \gamma_j$. As the clustering is separately done for every pattern $\gamma \in A$, in the following explanations for simplicity we consider only one element, $X \in \Gamma$ with $X = X^{(\gamma)}$, and X denotes the set of data for one pattern γ with $|X| = n$ and $X \subseteq [0,1]^p$ when using p distinct features.

4.1.1 Sample Grouping

Global features such as segmentation information and size are used to make rough assessments about the cluster nature. For example, in one of the applications, we used the number of handwriting strokes to divide the samples of a pattern into groups. Thus we can state that n samples of the overall data set X for one pattern γ_i are divided into a number of sample groups X_θ

$$\bigcup_{\theta \in \aleph} X_\theta = X ,$$

where θ denotes the ordering number of the inspected group. Through this data sample grouping step, we reduced the number of input samples from n to $|X_\theta| = n_\theta$ for the clustering process. However, we increased the number of clustering steps from 1 to $|\{\theta \in \aleph : n_\theta > 0\}|$ for each pattern class. Each data group within a pattern class is separately clustered to accomplish the rule generation process in step (3). The selection of a particular global feature for partitioning into groups is not mandatory. Depending on the application, suitable global features are employed.

4.1.2 Reduction of Clustering Features

After reducing the number of samples from n to n_θ in the first step, we reduce the number of features $|I|$ per sample in the second step. Typically, the number of local features describing a pattern is very high (130 in our application). These features are the clustering variables. In the mathematical sense, every feature is equivalent to one dimension in the feature space. The huge deviations of unimportant features for one pattern disturb the clustering convergence. Of course, an unimportant feature for one pattern can be of major importance for the next one. Therefore, the feature reduction can be locally done for each pattern and separately for each group of the specified pattern.

To make the clustering for each pattern more efficient, we select the major features for each pattern to be clustered. The feature selection process is based on two criteria. We select only those features that have high membership averages and also have low membership variances. The purpose of these two criteria is to select the feature with high discriminatory value. The remaining features with low membership values are used in the second part of the algorithm, to distinguish overlapping rules. The selection criteria are calculated with some statistics using the averages or means $\mu_{\theta,i}$ and standard deviations $\sigma_{\theta,i}$ for every feature $i \in I$ and $\theta \in \aleph$. For every $\theta \in \aleph$, we choose a subset $I_{\theta,i} \subseteq I$ of all features that will be considered in the following subsections. The reason for this constraint is the limitation of the training set used, i.e., the clustering of a few training samples relative to the large number of dimensions (features) would be rather ineffective.

4.1.3 Multiphased Clustering – the Process

In the third step, pattern groups $(X_\theta | \theta \in \aleph)$ of a pattern γ are clustered into a given number of clusters based on the selected features $I_{\theta,1}$. The input to the clustering algorithm is the data set $Y_1 = X_\theta | I_{\theta,1}$. The clustering is done with an established fuzzy clustering algorithm, for example, the fuzzy c-means algorithm or the Gustafson-Kessel algorithm. In the following discussion we will, for the sake of simplicity, describe the use of the fuzzy c-means algorithm [1].

We start the fuzzy c-means algorithm with $c = 2$. A natural maximum for c could be four or five. We denote the actual number of clusters as α in this iteration process. The result of one iteration are the centroids $z_{\theta,\alpha,j}$ of the α clusters and the corresponding membership values $m_{\theta,\alpha,j}(x)$ of the data $x \in Y_1$ to the clusters with J_α denoting the set of clusters and $j \in J_\alpha$. This means that every sample of the considered group of patterns Y_1 gets a membership value for every computed cluster.

The calculated centroids of the clusters represent the generated rules. But at this stage the number of features is still too high, so the clustering could have been partly ineffective. This implies that the number of features should be reduced again. This reduction is done by evaluating the data once again, with respect to the calculated centroids of the clusters. For this purpose we introduce a weighted deviation $w_{\theta,\alpha,i}(j)$. The weighted deviation is calculated for every considered feature $i \in I_{\theta,1}$. The standard deviation cannot be used in this case, because data has a membership value between zero and one for each cluster. The weighted deviation depends on the centroids $z_{\theta,\alpha,j}$ of the clusters, the membership values $m_{\theta,\alpha,j}(x)$ of the data to the clusters, and the data $x \in Y_1$ itself. Based on the weighted deviation, we select some features for each of the α clusters. These features represent the whole cluster from now on. The selection process is decided by the centroid of the cluster and the weighted mean of the membership value of each feature. The best few features for each cluster are considered to be representative of the rule generation process and are denoted by $I_{\theta,\alpha,j}$. This also reflects that the features that have many samples near the centroid are usually the best (high discrimination rate). Since the clustering was done for various $c = 2, \ldots, c_{max}$, the ideal α has to be found, where c_{max} is a user defined maximum for the number of clusters to be found.

The best cluster fit is found with the weighted deviation of the selected features $I_{\theta,\alpha,j}$ for each clustering step. The calculated mean of their weighted deviations for the various c clusters gives the measure of best fit. The cluster partition with the lowest deviation value is chosen as the solution. This decision criterion is deduced from one of the basics of clustering which states that for a good partition, the volume of a cluster should be as small as possible. For the chosen cluster number α_1, the best fit features are as follows:

$$I_{\theta,2} = \bigcup_{j \in J_{\alpha_1}} I_{\theta,\alpha_1,j}.$$

The first clustering phase was used to determine the most suitable features. This reflects the idea that we wanted to first concentrate on the individual properties of each pattern class. At this point in the algorithm, the number of ideal features $|I_{\theta,2}|$ is fixed, but the number of clusters α_2 is not. Therefore, an additional clustering process is required.

The c-means clustering is once again started from the beginning. The input to the clustering algorithm is the data set $Y_2 = X_\theta | I_{\theta,2}$. The clustering is done again for c = 2, ..., c_{max}. At the end of this new clustering step and estimation mechanism for the best few features and the best new clustering partition α_2, we obtain centroids, each with some features. This is the input to the rule base syntax generation step. In other applications it may be useful to introduce more clustering phases at this point.

4.1.4 Rule Generation in FOHDEL

The chosen cluster partition α_2 in step (3) (Figure 5) represents the number of rules found. The rules are deduced from the features selected out of $I_{\theta,2}$ of the centroid vectors. These features connected by the operator AND are the rule variables. Since each feature has a different membership value, attributes or linguistic terms have to be added to the feature variable. To estimate the quality of each feature, we calculate the histogram of the feature over the sample data. For features with a wide distribution covering several linguistic terms (like "between medium and high"), a combination of several attributes is possible. The rules are automatically generated in FOHDEL.

4.2 Rule Cross-Checking and Refinement

The rules have been independently generated for each pattern and for each group within a pattern thus far. For this reason, cross-checking of rules is required. Two types of conflicts can be identified. The first conflict arises, when two rules that are similar or mutually inclusive describe the same pattern. Such rules are named "redundant". Rule redundancy has no influence on the classification result, but has some influence on classification time. The second conflict may be that two rules that are mutually inclusive describe different patterns. In this case we say that the rules are "overlapped". This overlapping reflects a conflict in the rule base and affects the classification result. In order to resolve this second conflict, the rule base cross checking step is introduced.

Definition: If $r \in R$ be a rule in the rule base R, then we denote $\chi(r) \in A$ as the pattern which was the basis to formulate the rule in the first part of the algorithm. We call this the *corresponding pattern of the rule r*.

First, we have to clarify how a rule r has to be coded. Let $r \in R$ be a rule. The rule should then be coded as

$$r = \{(i, t_r(i)) \mid i \in I_r\} \wedge \forall (i, t_1), (i, t_2) \in r \Rightarrow t_1 = t_2$$

where $t_r(i)$ denotes the linguistic term that corresponds to the feature i for the rule r. It should be obvious that this is an adequate coding of the formulated rules, because the rules generated in the first part are conjunctively connected. In order to be able to define the various concepts of rule inclusions, we now define some relations.

Definition: Let the relation $\propto \subseteq \Sigma \times \Sigma$, with Σ denoting the set of all linguistic terms, $t_1, t_2 \in \Sigma$ and X being the universe of discourse, be defined as

$$t_1 \propto t_2 :\Leftrightarrow \forall x \in X : T(t_1)(x) \leq T(t_2)(x).$$

The relation can be extended to the following relation $\propto \, \subseteq R \times R$ as follows:

$$r_1 \propto r_2 :\Leftrightarrow \forall\left(i, t_{r_1}(i)\right) \in r_1 \exists\left(j, t_{r_2}(j)\right) \in r_2 :$$

$$i = j \wedge t_{r_1}(i) \propto t_{r_2}(j), r_1, r_2 \in R.$$

Rule r_1 is then said to be *included* in the rule r_2.

Now that we defined the concept of inclusion, we can define the aforementioned rule redundancy concept.

Definition: A rule $r \in R$ from a rule base R is said to be *redundant*, if and only if there exists another rule $r' \in R\backslash\{r\}$ in the rule base R, such that the following is valid:

$$r \propto r' \text{ and } \chi(r) = \chi(r').$$

After we have defined the term redundancy, we cannot define overlapping rules in a similar manner. Since we are working with a fuzzy rule base, a problem can arise. A rule that is not included in a different rule but is close enough to it, can fire both those rules while representing different patterns. Therefore, we first have to define the criterion for a linguistic term to be close to another one.

Definition: Let the relation $\sim \, \subseteq \Sigma \times \Sigma$ be defined as

$$t_1 \sim t_2 :\Leftrightarrow \exists x \in X : (\ (T(t_1)(x) = 1 \text{ and } T(t_2)(x) > 0\ ) \text{ or }$$
$$(T(t_2)(x) = 1 \text{ and } T(t_1)(x) > 0)\ ).$$

Now we introduce the concept of overlapping.

Definition: Let the relation $\prec \, \subseteq R \times R$ be defined as

$$r_1 \prec r_2 :\Leftrightarrow \forall\left(i, t_{r_1}(i)\right) \in r_1 \exists\left(j, t_{r_2}(j)\right) \in r_2 :$$

$$i = j \wedge t_{r_1}(i) \sim t_{r_2}(j), r_1, r_2 \in R.$$

A rule r_1 is said to be *overlapped* by another rule r_2, if and only if $r_1 \prec r_2$ and $\chi(r_1) \neq \chi(r_2)$.

If we consider that we have generated rules with r ∈ R (where R is the rule base), then the cross-checking operation has to compare every combination of two rules r_1 and r_2 where $r_1, r_2 \in R$. If the rule r_1 in the rule base is redundant, then we just remove r_1 from the rule base. By doing this we do not affect the results of the classification process, but just reduce the time needed to classify a pattern. If we find two rules r_1, $r_2 \in R$, where r_1 is overlapped by the rule r_2, then we have to change the rules to make them separable. Such overlapping is usually resolved by extending the rules. To find the best suitable features to extend the rules, statistical information is used. If the rules cannot be separated, then they are listed. To resolve this conflict, either an additional feature has to be added by the user or the pattern classification needs additional contextual interpretation (e.g., zero and letter o). The cross-checking process is repeated until either no overlapping exists in the rule base or the number of remaining conflicts cannot be reduced.

5   Word Classification with Context Dependence

Cursive script recognition systems utilize various levels of knowledge to help in recognition. The inherent ambiguity in handwriting presents the need for such knowledge in the system at word, syntax, sentence and semantics levels. These knowledge bases are used in contextual post-processing of results from the recognizer. At the first level, the recognizer may present possible sets of candidates for each character as output. Context knowledge is applied on this output to determine the possible strings that they can represent. At the next level, contextual knowledge at the sentence level can be used to determine which words are more likely to be correct in a given context. Knowledge bases at the syntax level can determine what words are more correct for grammatical validity of the surrounding text and at the semantic level, for the meaningfulness of phrases. These levels of context knowledge can hence be operative in producing a substitution set of possible outcomes of words as results in cursive script recognition.

The first and most common level of knowledge present is at the word level. This is used to determine the validity of a letter with respect to its surrounding characters. The system often presents a level of confidence or membership for each character alternative that it generates. These values can be used in computing a confidence value for the resulting candidate strings in the set that is generated. A thorough discussion of the other methods and strategies for word recognition using context information is reported by Lecolinet and Baret in [5].

A technique that overcomes the problems encountered by many common contextual recognition methods is that of lexicon look-up. Letter strings produced from the set of candidate sets of letters are checked against the lexicon to ensure that they are valid words and then accepted or rejected. When there is the possibility of more than one resulting word being valid, the membership values of the corresponding candidate letters that make up the word are used. A fuzzy look-up that presents all valid output words along with their fuzzy candidacy is then performed. The problem with any method that involves lexical look-up is that the representation of any reasonably realistic vocabulary will involve a large dictionary, with consequently large memory and search time requirements. This trade-off has to be considered with respect to the end-use of the recognizer and the degree of precision required.

6   Application and Concluding Discussions

We combined the previously introduced theoretical models into an automatic handwriting recognition system (see Figure 6). The application is limited to postal address recognition, which includes ZIP codes, town names, country names and other address features. The input to the system is off-line handwriting data acquired from NIST®, CEDAR® and Siemens Electrocom®.

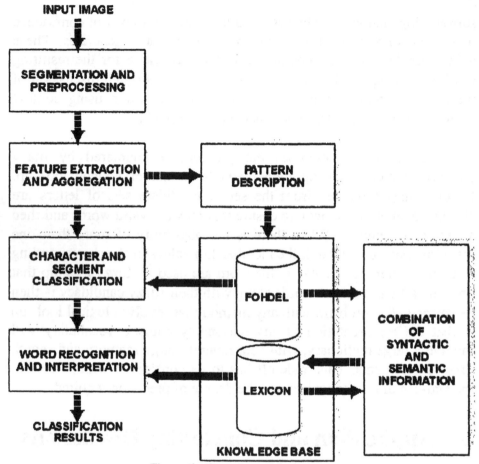

Figure 6. Overview of the system.

The classification is accomplished through parallel evaluation of all rules contained in the rule base. We, therefore, first segment the word into different groups and find characters or syllables, and then evaluate them against the rule base. The evaluation specifies the fuzzy membership values of the presented patterns (usually single characters) to the different possible pattern classes described by various rules.

In order to accelerate the classification process and increase the classification rate of the character recognition process, we introduced a low-level preclassification step. This is based on global information, and is done in what we called the Spitz-Coding step. We thereby extended the concept of WSTs (Word Shape Tokens) that Spitz [17] used for recognizing printed words. We then inspect the surrounding box of one character and its relationship to the surrounding boxes of other characters and the whole word itself. By doing this we deduce the class of characters to which the specified character belongs. Given below are some examples of classes of lower case letters that we defined to be an extension of those outlined by Spitz. (1) The character class containing only the character f, the class with lower case letters going both under the base line and above the center line. (2) The character class containing g, j, p, q and y, the class with letters going under the base-line. (3) The character class containing the b, d, h, k, l and t, the class of lower case letters going above the center line. (4) The class with the rest of the lower case characters, those that are written between the base line and the center line. To use this Spitz-Code, we had to introduce various preprocessing steps performed on the input data, like a base-line correction step.

By means of individual character recognition, we succeeded in recognizing parts of words. The character recognition step uses rules generated by the automatic rule base generation algorithm described in Section 4 and based on linguistic modeling with the help of fuzzy logic introduced in Section 2 and the feature extraction methods described in Section 3. By performing the fuzzy look-up procedure introduced in Section 5, we significantly reduced the number of possible words. Successive recognition of previously unrecognized parts of the word recursively reduces the number of possible solutions. There are different ways to accomplish this. One possible method is to split one group of segments into two or more, and try to recognize characters in the new groups. The other method is to combine two or more groups into one, and similarly attempt character recognition. Yet another very effective way to recognize whole words is to use verification techniques instead of recognition techniques. In this method, we match every entry in the reduced lexicon with the original or preprocessed picture. In other words, we try to verify whether the word that is to be read is a word contained in the reduced lexicon. The disadvantage of this is that it is a time-consuming calculation, because it is proportional to the number of entries in the reduced lexicon. Therefore, in real-time applications we have to reduce the entries in the lexicon as much as possible through recognition techniques. The verification task is not as difficult to perform as the recognition task because one picture and one possible solution are considered at a time. Therefore, in applications where time is not a crucial factor, this verification technique is a promising new idea.

Figure 7 shows the results of intermediate stages of the proposed methodology. First, an input image from a postal address of an envelope as the region of interest for city name recognition is located and binarized. The next stage shows the result of our preprocessing and segmentation stage with labeled segments. The third image shows the surrounding boxes of characters or segments that are used for the approximate assessment of Spitz-Code. The results of the character recognition phase are shown on the right hand side including the rule numbers in the knowledge base and the corresponding membership values in the universe of discourse [0,100]. The final classification of the given example is the city "Krefeld" with 85% possibility from the given lexicon of German town names.

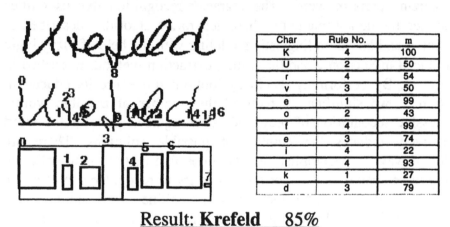

Char	Rule No.	m
K	4	100
U	2	50
r	4	54
v	3	50
e	1	99
o	2	43
f	4	99
e	3	74
i	4	22
l	4	93
k	1	27
d	3	79

Result: **Krefeld**   85%

Figure 7. Steps of word recognition.

In this chapter we outlined the basic idea of fuzzy word recognition with context dependent rules. These rules are automatically generated and coded in a dedicated language, which describes the characteristic features of the words. The detailed description of various recognition modules (Figure 6) can be found in the referred works. The classification results showed an error rate of approximately 15 to 20% without lexical analysis and approximately 8 to 9% with the lexical analysis. A detailed analysis of the classification benchmarks could not be produced because of lack of data with ground truth information for verification. The scope of this method is not limited to the off-line handwriting recognition, but it is extendable to other similar applications. The presented idea will be applied in our robotics lab for doorplate recognition and an in-house mail processing and distribution system. The fusion of syntactic and semantic knowledge is the biggest challenge, and we expect more useful algorithms and applications evolving from the presented methods.

Acknowledgments

The work presented here is part of the READ-project, which is supported by the German Ministry of Science and Technology [12]. We are thankful to Duy Tran, Volker Mueller, Dr. Jayati Ghoshal and Dr. Greg Watkins and for their cooperation and help.

References

[1] Bezdek, J.C. (1981), *Pattern Recognition with Fuzzy Objective Function Algorithms*, Plenum Press, New York.

[2] Fu, K.S. (1982), *Syntactic Pattern Recognition and Applications*, Prentice-Hall, NJ.

[3] Ivancic, F., Malaviya, A. and Peters, L. (1998), "An Automatic Rule Base Generation Method for Fuzzy Pattern Recognition with Multiphased Clustering," *2nd International Conference on Knowledge-Based Electronic Systems*, Adelaide.

[4] Klir, G. and Yuan, B. (1995), *Fuzzy Sets and Fuzzy Logic*, Prentice-Hall, NJ.

[5] Lecolinet, E. and Baret, O. (1994), "Cursive Word Recognition: Methods and Strategies," in S. Impedovo (Ed.), *Fundamentals of Handwriting Recognition, NATO ASI Series F: Computer and Systems Sciences*, Vol. 124, pp. 235-263, Springer Verlag.

[6] Lee, C.C. (1990), "Fuzzy Logic in Control Systems: Fuzzy Logic Controller," *IEEE Transactions on Systems, Man, and Cybernetics*, Vol. 20, No. 2, pp. 404-435.

[7] Malaviya, A. (1996), *On-Line Handwriting Recognition with a Fuzzy Feature Description Language*, GMD-Report Nr. 271, Oldenbourg Verlag, ISBN 3-486-24072-2, Munich.

[8] Malaviya, A. and Klette, R. (1996), "A Fuzzy Syntactic Method for On-line Handwriting Recognition," in Perner, Wang and Rosenfeld (Eds.), *Lecture Notes in Computer Science* 1121, Springer Verlag.

[9] Malaviya, A., Leja, C. and Peters, L. (1997), "A Hybrid Approach of Automatic Fuzzy Rule Generation for Handwriting Recognition," Downtown and Impedevo (Eds.), *Progress in Handwriting Recognition*, World Scientific, Singapore.

[10] Malaviya, A. and Peters, L. (1995), "Extracting Meaningful Handwriting Features with Fuzzy Aggregation Method," *IEEE Proceedings of ICDAR '95*, Montreal.

[11] Malaviya, A. and Peters, L. (1997), "Fuzzy Feature Description of Handwriting Patterns," *Pattern Recognition*, Vol. 30, No. 10, pp. 1591-1604.

[12] Malaviya, A. and Peters, L. (1998), "Read: Document Analysis and Recognition System," *ERCIM News Letter*, Paris.

[13] Malaviya, A., Peters, L. and Theissinger, M. (1994), "FOHDEL: A New Fuzzy Language for On-Line Handwriting Recognition," *FUZZ-IEEE*, pp. 624-629, Orlando.

[14] Peters, L., Leja, C. and Malaviya, A. (1998), "A Fuzzy Statistical Rule Generation Method for Handwriting Recognition," *Expert Systems*, Vol. 15, No.1.

[15] Rosenfeld, A. (1984), "The Fuzzy Geometry of Image Subsets," *Pattern Recognition Lett.* 2, pp. 311-317.

[16] Shaw, A.C. (1969), "A Formal Picture Description Scheme as a Basis for Picture Processing Systems," *Information and Control*, Vol. 14, pp. 9-52.

[17] Spitz, A.L. (1997), "Moby Dick meets GEORC: Lexical Considerations in Word Recognition," *Proceedings of 4th International Conference on Document Analysis and Recognition*, Vol. 1, Ulm.

[18] Watkins, G. (1997), *A Framework for Interpreting Noisy, Two dimensional Images, based on a Fuzzification of Programmed, Attributed Graph Grammars*, Ph.D. Thesis, Rhodes University.

[19] Yager, R.R. and Zadeh, L.A. (1992), *An introduction to fuzzy logic applications in intelligent systems*, Kluwer Academic, Boston.

[20] Yau, K.C. and Fu, K.S. (1979), "A Syntactic Approach to Shape Recognition Using Attributed Grammars," *IEEE-SMC*, Vol. 9, No. 6, pp. 334-345.

Chapter 11:

License Plate Recognition

Literature-Based Recognition

LICENSE PLATE RECOGNITION

M.H. ter Brugge, J.A.G. Nijhuis,
L. Spaanenburg, and **J.H. Stevens**
Department of Mathematics and Computing Science
Rijksuniversiteit Groningen
P.O. Box 9700, 9700 AV Groningen
The Netherlands

Real time license plate recognition is a key function for automated traffic monitoring and law enforcement on public roads. In the VIPUR system for Vehicle Identification on PUblic Roads, this function is mainly created by neural engineering. The current implementation provides all steps from image collection to plate understanding. A hierarchical structure of experts is designed in which various techniques like shift invariant template matchers, neural networks and LVQs are applied. Together with recursive segmentation and the use of classifier combining methods, a real life recognition rate of 85% with less than 0.001% false acceptance is achieved.

1   Introduction

The automatic identification of vehicles by the contents of its license plate is important in private transport applications such as travel time measurements [1], parking lot traffic management [2], toll collection [3], speed limit enforcement [4], and identification of stolen cars [5]. Especially the use of a Car License Plate Recognition (CLPR) system in a real time/life environment puts high demands on the throughput of the system and the contents of the image. The use as a law enforcement system requires that not only the license plate but also a large (recognizable) part of the car is visible in the image; therefore, the license plate occupies only a small portion of the overall image. Further, a low cost, easy to replace camera is preferred, so the plate image will be of a low resolution.

Figure 1. Some input examples for the CLPR system. The images are recorded by a police-operated monitoring system. Each image contains 768×567 pixels with 8-bit gray scale information per pixel.

One may conclude from Figure 1 that a fixed location for the license plate within the overall image cannot be assumed. Virtually each pixel may belong to a license plate and, therefore, needs to be processed. In order to meet the throughput requirements, it makes sense to reduce the area of interest as soon as possible. Only those parts of the input image that fulfill a set of license plate properties need to be considered for further inspection.

The limited resolution of the recorded characters together with dirt (dust, flies), screws and bolts (used to attach the plate to the car), overhanging car parts (tow bar mounted too high), etc. make the development of a reliable optical character recognizer (OCR) very complicated. Many of the approaches common in conventional OCR systems [6] turn out to be useless. It is therefore necessary to build a system that is capable of exploiting both rules defined by the license plate registration regulation (explicit knowledge) [7] and features present in the measurement of the OCR task on hand (implicit knowledge) [8].

The design of the CLPR system for Vehicle Identification on PUblic Roads (VIPUR) is based on the Dutch license plate registration regulation. This regulation enholds a strict set of rules for the position of the license plate on the car, the shape, size, color of the plate, the character font (see Figure 2), etc.

The first VIPUR release [9] focused largely on the exploitation of fuzzy logic to reason from the registration regulation. In [10] the character recognition is pushed onward. Such dedicated systems are hard to tune and maintain, because little of the software code can be reused. A more formal specification to generate the different pieces of code by the use of DT-CNNs is introduced in [11]. Here we overview and extend this work with mixed expert systems for implementation, while retaining the impressive high recognition rate.

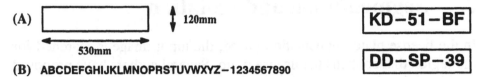

(A) 120mm   530mm

KD–51–BF

(B) ABCDEFGHIJKLMNOPRSTUVWXYZ–1234567890

DD–SP–39

Figure 2. License plates with black characters on a yellow retroflexive background according to model 18.1 (A) with characters according to model B.1 (B) [RVW 6711]. The height of the characters varies from 75mm to 77mm (10mm for the horizontal bar). The width (margins included) varies from 32mm to 87mm. The characters are symmetrically located on the license plate.

The CLPR system (Figure 3) consists of four main units: a segmentation unit, an isolation unit, a recognizer and a syntactical analyzer. The segmentation unit is completely implemented by DT-CNNs [12] and determines the location of the license plate based on structural features and some weak size constraints. The extracted plate is passed to the character isolator that tries to locate the characters on the plate. This module is almost completely implemented by DT-CNNs. The isolated characters are processed by the recognizer. This part of the CLPR system uses multilayer perceptrons (MLPs) and template matchers. The syntactical analyzer checks whether the candidate characters returned by the recognizer satisfy a number of syntactic rules that exist for Dutch license plates. If these rules are not satisfied, or one of the characters is unrecognizable, the image is rejected.

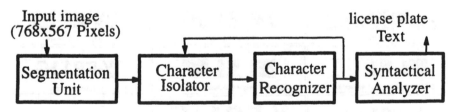

Figure 3. Outline of the Car license plate Recognition System.

Section 2 discusses the segmentation of the license plate and the isolation of individual characters. Then we focus our attention on feature-based classifiers, followed in Section 4 by other classification schemes. Finally, in Section 5, we introduce the overall system and compare VIPUR to other systems.

2   Segmentation and Isolation

In the license plate segmentation phase, the input image is searched for the region that contains the license plate. Several authors have proposed solutions for this problem [5], [10], [13], [14]. All of these methods incorporate a number of features of the license plate; for example, "a license plate is a rectangular area on the car, which contains a number of dark characters." Inspired by this kind of explicit knowledge, two features, grayness and texture, have been appointed to each pixel in the image. In [10] fuzzy membership values between 0 and 1 were used for both features. In [11] DT-CNNs are used, which forces us to represent grayness and texture by crisp values -1 and 1. A grayness of 1 indicates that the pixel has a gray value that corresponds to the color of a license plate. A grayness of -1 indicates that the color of the pixel is outside the range of gray values that is common for license plates.

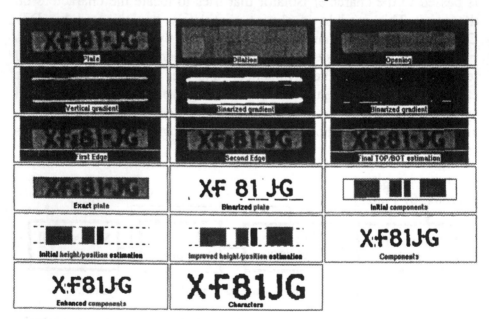

Figure 4. The segmentation of the license plate consists of several steps: (1) the exact location and orientation of the license plate is estimated (shown in the first three rows), (2) the license plate is binarized, (3) initial connected components are used as candidate characters and (4) final segmentation into isolated characters.

The segmentation on grayness and texture leaves a number of locations for potentially correct license plates. This number is reduced by applying some common sense with respect to the size of the plates. After further binarization, the characters within the plate can be isolated. The plate with the highest success in isolating characters is assumed to be the license plate that has to be recognized. Figure 4 illustrates the overall process of segmentation and isolation.

2.1   Soft Segmentation of License Plates

The ranges for grayness and texture are determined by a histogram-based method. For the grayness feature, the gray values of pixels taken from a large number of exemplary license plates are used to construct a frequency table (the number of occurrences of each gray value in each license plate). Based on this table, the value range is derived. It turns out that the gray value of most license plate pixels are in the range [0.1, 0.57]. For the texture feature the histogram is constructed by applying a 3×3 Sobel operator to each license plate pixel. It turns out that most license plate pixels have an absolute Sobel value larger than 0.73.

The range for grayness has an upper bound and a lower bound. This requires a two-step DT-CNN evaluation. In the first step, all pixels of the input image (Figure 5a) with a gray value larger than 0.1 are extracted (Figure 5b). In the second step, all pixels with a gray value larger than 0.57 are deleted from the output (Figure 5c).

Note that not all pixels on the plate are black in Figure 5c. Particularly, this is the case for pixels that are part of a character. Such pixels are not selected since their gray value is too low. One way to solve this problem is reducing the lower bound of the grayness range. However, this makes the grayness feature worthless. Another way to solve the problem is to make all white pixels black if they are close to one or more black pixels. Morphologically, such an operation is described by a dilation. We have chosen a 7×7 block structuring element to dilate the grayness image (Figure 5d). The choice of the structuring element is based on the assumption that each character consists of a set of narrow lines that have width of less than 8 pixels. According to [15] this dilation is implemented by a DT-CNN with a 7×7 template. To avoid the use of templates larger than 3×3, the 7×7 structuring element is decomposed into three successive dilations with a 3×3 block structuring element before a mapping onto DT-CNNs is constructed [16].

For texture evaluation the Sobel operator is mapped onto a DT-CNN. In [15], [17], [18], it is shown that morphological operators can be mapped onto DT-CNNs. For traditional filter operations (possibly combined with traditional set operations), a similar mapping can be constructed. For example, the combination of a filter operation with a mask M and a threshold operation with threshold T is implemented by a DT-CNN with a control template that is equivalent to the transpose of M, a zero feedback, and a bias that is equivalent to -T. This rule is used to derive the template for the first step, in which all pixels with a Sobel value larger than 0.73 (dark/light transitions) are extracted. In the second step, all pixels with a Sobel value smaller than -0.73 (light/dark transitions) are added (Figure 5e).

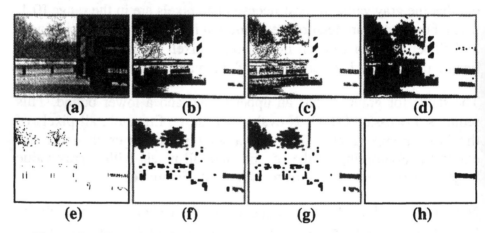

Figure 5.   Extraction of potential license plates: (a) is the input image, (b) indicates the pixels that are bright enough, (c) is the grayness feature, (d) is the dilated grayness feature, (e) is the texture feature, (f) is the dilated texture feature, (g) is the combination of texture and grayness, and (h) is the image that contains all plates after deleting undersized and oversized plates.

Notice that for edge/texture detection, we have constructed a two-layer Sobel DT-CNN instead of using the traditional single-layer Laplace DT-CNN. The disadvantage of Laplace is that it is a second-derivative operator, which is unacceptably sensitive to noise (which is always present in our case). Therefore, Laplace is hardly ever used by the image processing community for edge detection.

Similar to the initial grayness image, the texture image also needs to be dilated. For this purpose we use a 15×9 block structuring element, which can be decomposed into four 3×3 block structuring elements and three 3×1 block structuring elements. The reason for not mapping the last dilation is that it can be combined with the following operation. Remember that a part of the image is very likely to be a license plate if it has the appropriate grayness and the appropriate texture. The combined feature is obtained by computing the intersection of the grayness and the texture image (Figure 5g). The grayness image is stored on the input of the second DT-CNN. The next step this DT-CNN has to perform is dilating the output (the postponed dilation) and computing its intersection with the input image.

2.2 Introducing Size Constraints

Notice that Figure 5g still contains a number of objects that are not license plates. Apparently, objects like trees and parts of the crash barrier have the same characteristics as license plates. To further reduce the number of nonplate components, we use a number of weak size constraints as described below. After applying these constraints, the image in Figure 5h is obtained. Typically, we lose around 5% of the plates due to segmentation problems.

The four weak size constraints are minimum/maximum height and minimum/maximum width. The width of a plate is at least 91 pixels while the height is at least 13 pixels. The rejection of smaller plates at this stage turns out to have no effect on the overall performance of the system, since these plates are unreadable anyway due to the limited character resolution. The maximum width of a plate is 171 pixels and the maximum height is 61. The size constraints can be computed in parallel by four DT-CNNs. First, a selector image is computed for each size constraint (Figure 6a-d). Such an image contains pixels at positions where a horizontal or vertical bar of a certain length fits in the image. For example, the minimum height selector contains a pixel everywhere a vertical bar of height 13 fits in the image. Morphologically, a size selector is found by means of the erosion operator with the appropriate structuring element. The minimum height selector is found by an erosion with a 1×13 block structuring element. Similar to dilation, erosion with a 1×13 block structuring element is decomposed into six erosions with a 1×3 block structuring element.

The first step dilates the image that is stored on the input (Figure 5g), while the following five steps perform the additional dilations on the output. After computing the selection, all objects need to be restored (Figure 6e). For this purpose each object in the output is enlarged (dilated with a 3×3 cross structuring element) step by step. In order to avoid adding pixels that are not part of the object in the original image (that is still on the input), an intersection with the input is required after each step.

The three other selectors are restored in an identical way (Figure 6f-h). Finally, restoration results need to be combined. The set of objects that are large enough to be a license plate is found by computing the intersection of Figure 6e and Figure 6g. Then, all objects that are too large are deleted by subtraction (using Figure 6f and Figure 6h). The three combination operations (one union and two set subtractions) can be computed by a three-step DT-CNN [18].

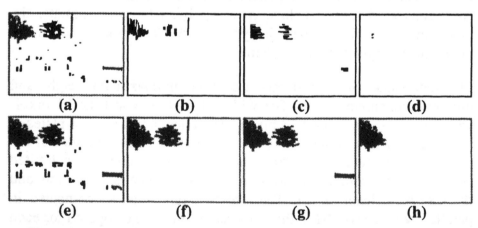

Figure 6. Processing size constraints: (a) is the selector for objects that satisfy the minimum height condition, (b) is the selector for objects that do not satisfy the maximum height condition, (c) is the minimum width selector, (d) is the maximum width selector, (e)-(h) are the corresponding reconstructions.

2.3 Binarization

An image is made available for further processing by *binarization*: the picture is taken, converted to gray scale and coded by digital values for the pixels. Then, the individual regions of interest are labeled by segmentation before recognition is performed. Most processing systems assume the use of a high quality raster image; however, in license plate recognition the quality of the camera will not be high and the region of interest is only a small portion of the total image. The processable image is therefore of low quality with variable background intensity, low contrast and stochastic noise.

Two classes of binarization methods are apparent from literature [19]. *Global binarization* calculates a single threshold value for the entire image. Pixels having a gray level darker than the threshold value are labeled black, otherwise white. In contrast, *local methods* compute a threshold for each pixel in the image on the basis of information contained in a neighborhood of the pixel. Some methods even calculate a threshold surface.

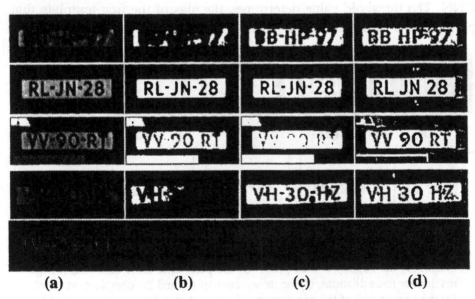

(a)	**(b)**	**(c)**	**(d)**

Figure 7. The difference between various thresholding schemes: (a) the original image, (b) global, nonadaptive fixed threshold value, (c) global adaptive, and (d) local adaptive.

It is essential to provide a binarization, which will correctly label all the information present, even in low contrast areas while being insensitive to variable background intensity and stochastic noise. Therefore, we have verified the results of [19] for our specific problem. Some of the results are visualized in Figure 7.

The segmented license plates are clearly of varying quality. The influence of the background intensity becomes apparent when applying a straight nonadaptive binarization: 2 of the 5 plates become even less readable. When trying to remedy this by an adaptive threshold, other plates become unreadable. It is clear from the Figure 7 that local binarization with an adaptive threshold appears most promising: the readability of all plates has increased. This observation supports the limited practical evidence given in [19].

Binarization is performed as part of the second system module, responsible for the isolation of the individual characters on the license plate. This module is implemented by a single DT-CNN with a time varying template, and a very small part of traditional programming. First, a dynamic threshold is determined using traditional methods [10]. Then the gray scale image (Figure 8a) is stored on the input of the DT-CNN. The threshold value determines the bias of the first template that performs the actual threshold operation (Figure 8b).

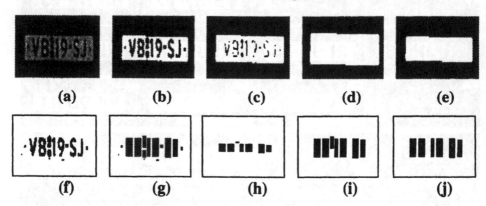

Figure 8. Isolation of characters: (a) the original image, (b) the binarized image, (c) the slightly eroded plate, (d) the plate after three erosions, (e) the result after three dilations, (f) the intersection of (b) and the complement of (e), (g) the convex hull of the components, (h) the selector for components that are large enough, (i) the set of components that are large enough, and (j) the extracted characters.

2.4 Character Isolation

Before performing a connected component search, the black area around the plate is removed. For this purpose we first determine what this area is (Figure 8e). Since characters are thin objects, characters can be deleted from the image by a so-called morphological opening. This is an erosion with a certain structuring element followed by a dilation with the same structuring element. Again, the choice of the structuring element (a 7×7 block structuring element) depends on the thickness of the characters.

The network output after time step 2 and time step 4 are shown in Figure 8c and Figure 8d, respectively. In time step 8, the components on the plate are extracted by subtracting the output from the binarized input image, as shown in Figure 8f. After that, a rectangular hull extraction is performed (Figure 8g). During this iterated process, white pixels that have at least two black 4-neighbors are painted black each step, until the network converges.

Finally, a minimum height criterion similar to one of the criteria used for license plate segmentation is applied. Since the minimum character height is 11 pixels, a height selector image is evaluated by five successive erosions with a 1×3 block structuring element (Figure 8h). Restoration of the objects is much easier than the restoration operation that is described in the previous section. Since all objects are rectangles, restoration is done by five dilations with a 1×3 block structuring element (Figure 8i). Removal of boxes with a height less than 11 pixels is therefore performed by following the ten time steps.

The last step before characters are cut out and passed to the recognizer is a selection of objects with almost equal height. Since it is very hard (and probably impossible) to do this by means of DT-CNNs, this last step is performed using traditional programming. Finally, the remaining set of objects is used to cut out the characters from the original plate (Figure 8j).

The selected components are passed on to the recognizer module only if they make up a valid license plate. A valid license plate contains six selected components that are all of the same height and start at the same vertical position. In all other cases the license plate is marked by the system as unrecognizable and therefore rejected. The current systems reject about 6% of all images during the segmentation/isolation stage.

3   Feature-Based Recognition

In order to classify characters, it is required that the isolated images are quantified such that a recognizer can see the difference. Presenting the image directly to the recognizer makes it hard to handle transformed images (resulting from rotation, sizing and/or translation). Therefore, it is necessary to derive a set of features that are invariant to such transformations. From literature, one finds three techniques aiming at

- moments that represent the composition,

- principal components that reflect the composition,

- side-way projections of the composition.

In the following sections, we will present these techniques and show that a combination of these techniques produces good results.

3.1   Moments

Moments are numeric characteristics of a distributed set of objects, measured relative to a point of reference. They appear in various disguises, such as in stochastics and mechanics. In a gray-toned image, moments are based on n pixels P_i with a gray level m_i and a distance r_i:

$$moment = \sum_{i=0}^{n} (m_i \times r_i) \tag{1}$$

In other words, a moment gives a measure of the internal structure. If the point of reference (the origin) is chosen at the heart of the image, the measure is called the *central moment*. In this way, the moment has become invariant to the rotation and sizing of the segmented part. The determination of this mass origin for two character sets is shown in Figure 9.

The impact of the different character parts on the moment can be differentiated by raising the mass and distance contributions within the formula to, respectively, the powers p and q:

$$u_{pq} = (p,q) - order\ moment = \sum_{i=0}^{n} (m_i^p \times r_i^q) \tag{2}$$

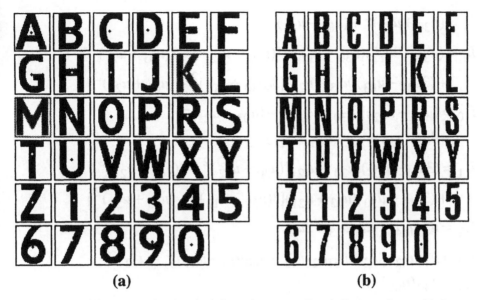

Figure 9. The two main character fonts in use on Dutch license plates, (a) the normal font and (b) a smaller "American" font. The cross indicates the mass center of each character.

A large range of (p,q)-order moments can be generated and, therefore, it is of interest to find which combination will actually provide compound moments (features) that support the recognition of the different characters [20]. Lengthy analysis brought us to the following set:

$$I_1 = u_{20}u_{02} - u_{11}^2$$

$$I_2 = (u_{30}u_{03} - u_{21}u_{12})^2 - 4(u_{30}u_{12} - u_{21}^2)(u_{21}u_{03} - u_{12}^2)$$

$$I_3 = u_{20}(u_{21}u_{03} - u_{12}^2) - u_{11}(u_{30}u_{03} - u_{21}u_{12}) + u_{02}(u_{30}u_{12} - u_{21}^2)$$

$$I_4 = u_{30}^2 u_{02}^3 - 6u_{30}u_{21}u_{11}u_{02}^2 + 6u_{30}u_{12}u_{02}(u_{11}^2 - u_{20}u_{02})$$
$$+ u_{30}u_{03}(6u_{20}u_{11}u_{02} - 8u_{11}^3) + 9u_{21}^2 u_{20}u_{02}^2$$
$$- 18u_{21}u_{12}u_{20}u_{11}u_{02} + 6u_{21}u_{03}u_{20}(2u_{11}^2 - u_{20}u_{02})$$
$$+ 9u_{12}^2 u_{20}^2 u_{02} - 6u_{12}u_{03}u_{11}u_{20}^2 + u_{03}^2 u_{20}^3$$

$$I_5 = u_{20} + u_{02}$$

$$I_6 = (u_{20} - u_{02})^2 + 4u_{11}^2$$

$$I_7 = (u_{30} - 3u_{12})^2 + (u_{03} - 3u_{21})^2$$

$$I_8 = (u_{30} + u_{12})^2 + (u_{03} - u_{21})^2$$

$$I_9 = (u_{30} - 3u_{12})(u_{30} + u_{12}) \times [(u_{30} + u_{12})^2 - 3(u_{03} + u_{21})^2]$$
$$+ (3u_{21} - u_{03})(u_{03} + u_{21})[3(u_{30} + u_{12})^2 - (u_{03} + u_{21})^2]$$

$$I_{10} = (u_{20} - u_{02})[(u_{30} + u_{12})^2 - (u_{03} + u_{21})^2]$$
$$+ 4u_{11}(u_{30} + u_{12})(u_{03} + u_{21})$$

$$I_{11} = (3u_{21} - u_{03})(u_{30} + u_{12})[(u_{30} + u_{12})^2 - 3(u_{03} + u_{21})^2]$$
$$+ (3u_{12} - u_{30})(u_{03} + u_{21})[3(u_{30} + u_{12})^2 - (u_{03} + u_{21})^2]$$

$$I_{12} = u_{40}u_{04} - 4u_{31}u_{13} + 3u_{22}^2$$

$$I_{13} = u_{40}u_{22}u_{04} - 2u_{31}u_{22}u_{13} - u_{40}^2 u_{13} - u_{04}u_{31}^2 - u_{22}^3)$$

$$I_{14} = I_4 / (u_{00}I_2)$$

$$I_{15} = I_1^2 / (u_{00}I_3)$$

$$I_{16} = I_1 I_3 / I_4$$

Features I_1 to I_4 are almost universally cited to be suitable [21] and already provide a 60% recognition. From [22] we use the invariants I_5 to I_{11}; however, we experienced no need for normalization of the central moments, although suggested in the cited paper. Features I_{12} and I_{13} are again taken from [21]. As some characters (such as 6 and 9) are hard to recognize by rotation-invariant moments, we added I_{14} to I_{16}. With these features, a multilayer perceptron can be trained. The 16 features are presented to 16 input neurons. The network is best with 30 hidden neurons, providing a 90% recognition; with more than 30 hidden neurons, the network becomes over-parametrized. As the errors are largely false accepts, a moment-based classifier seems not individually usable but requires combination with classifiers of another type.

3.2   Principal Component Analysis (PCA)

Principal component analysis [23] is a statistical method which determines an optimal linear transformation y = Wx for a given input vector x of a stationary stochastic process and a specified dimension m of the output vector y, where w is the desired transformation matrix. In pattern recognition and communication theory, the PCA is known as the Karhunen-Loéve or Hotelling transform. The intention is to compress the information contained in the input data (i.e., m<<n) by exploiting statistical regularities. In fact, PCA transforms correlated input data into a set of m statically decorrelated features (or components), usually ordered according to decreasing information content.

The first principal component y_1 is the normalized linear combination of the components of the input data vector which has, on average, the largest output variance. Similarly, the second principal component y_2 has, on an average, the largest variance among the directions orthogonal to the direction of y_1. Then the third principal component y_3 is taken in the maximum variance direction in the subspace perpendicular to the first two eigen vectors, and so on. Although seemingly computationally elegant, the standard approach that uses a precomputed autocorrelation matrix is impractical for large m. The alternative is an adaptive learning algorithm for on-line estimation of the eigen vectors directly from the input data.

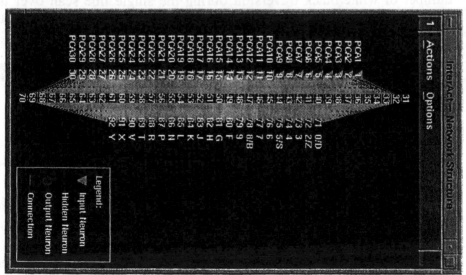

Figure 10. Neural network classifier with PCA input features.

The vector is constructed from moving with a "looking glass" over the picture. Each element of the vector reflects the local content, either in terms of contour or of bending points, as one of a set of primitive patterns. The image is sliced in four directions (horizontal, vertical, diagonal and off-diagonal) and of a predetermined width. A feature is then given by the number of primitive patterns of the same type in each slice. Using PCA, a feature set can be composed of the minimum amount of features that allow the individual characters to be recognized. In our case, we use the first 30 components of the Karhunen-Loéve transform (Figure 10), as using more components gave rise to noticeable unlearning effects, probably caused by the fact that with increasing components, the amount of stochastical noise in the rest class also increases.

3.3   Projection

A third way to quantify characters is by sideways projection [24]. A straightforward way to accomplish this is by one-dimensional counting of black pixels. This can be viewed to be related to texture analysis, where next to this simple pixel-based information, one soon takes recourse to runs of similar pixels. An alternative could therefore be the distribution of runs of a specific length in the character image. Most popular is the Connected Component Count, where one counts the number of white/black traversion during the single line scan over the image. This latter technique can be easily realized by a DT-CNN (Figure 11).

Figure 11. An example of the vertical connected component extraction by a DT-CNN. Each dot captured in the screen corresponds to a neuron in the DT-CNN. The vertical connected component count can be read after the DT-CNN is converged as shown in the last screen shot.

By means of DT-CNNs, four different features are generated: horizontal projection, vertical projection, horizontal connected component count, and vertical connected component count. For each feature, a specific DT-CNN has been constructed that is fed with a normalized version (the height of each character is scaled to 15 pixels; the width is scaled proportional to the height) of the binary connected components.

.4 Combining Classifiers

\s shown from the above discussion, feature-based neural classifiers
/ill not be perfect. Neither of them will achieve a 0% error at an
cceptable recognition rate. This is caused by the imperfection of the
mited feature sets and the balance between false and correct
cceptance. When a neural network is not permitted to make false
ccepts, the amount of rejected plates increases considerably (Figure
2).

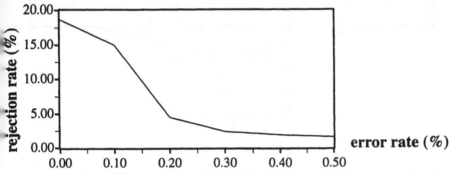

Figure 12. The error-reject curve for one of the neural network classifiers.
Without rejects this network correctly recognizes 98.97% of all input patterns.

.s a consequence, it is of interest to see whether or not the false
ccepts of the different networks are so much different that a
ombination of the networks would allow more perfect decisions
Figure 13). As the feature-based classifiers use neural networks, they
in handle only characters that appear often on license plates. In every
ountry, some characters are taken out for special purposes and this
akes their occurrence so rare that neural training cannot be
alistically performed. Further, to increase the recognition reliability,
ich net is followed by a security check to determine per character
ass the absolute value for the winner and the runner-up. This
etermines the dependability of the recognition.

he voting scheme is based on a class selective error reject. As 70% of
e cut-outs are sufficiently recognized by all nets, they can be labeled
 classified. When no net is able to recognize the cut-out, they can be
jected. All other cut-outs have to be inspected more closely by
bsequent means.

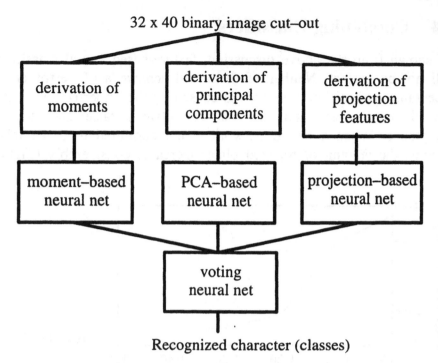

Figure 13. Schematic overview of the recognition module. Multiple neural network classifiers, each operating on a different set of input features are used in parallel. A confusion-based voting scheme produces the final classification result for all neural networks.

4   Other Means of Recognition

Thus far, we have refrained from a classification directly on the cut-ou i.e., without feature preprocessing. Such an approach will be optim [2], [9], [25] but consumes too much time and storage space when th accuracy requirements are high. Nevertheless, their usage on small subproblems can be of advantage. Therefore, in the following we wi study such classifiers and conclude with the overall architecture of th license plate recognition system VIPUR.

4.1   Template Matching (LVQ)

License plates contain printed characters of a well-defined font. Th fact that the plate can be bent, screws and bolts can distort the vie and light may be diffused pose problems additional to the fact that low cost camera is used to observe a vehicle moving at unknown spee

In other words, the actual characters are not only translated, scaled and rotated in the view, but also distorted and are of low precision. Template matching aims to find a direct relation between ideal and actual character by simply comparing normalized representations of both.

The idea is that the various characters can be clustered, such that all instances fall inside a cluster and no clusters overlap. The distance can be measured from a presented cut-out to all clustered prototypes and this distance then expresses a degree of alikeness (Figure 14). Thirty prototypes are used for training the clusters, different in the overall impact of the disformations, but characteristic for all practical situations.

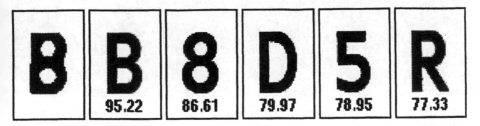

Figure 14. The 5 best matching templates with the score for the presented cut-out character "B."

An attempt to cover all variations of all characters will simply be too much for comfort. Instead, we have largely implemented

- dedicated template matchers for single symbols and/or classes with highest accuracy;

- a general-purpose template matcher with only a few representative examples for a formal check on an existing classification.

It may be clear that some matching scores enable a direct classification while others are more hesitant. Where a full-size dedicated template matcher might have enough discriminatory information to perform a reliable classification, this is not true for the general-purpose matcher. But here a Bayes decision can be made by the preknowledge that emanates from preprocessing by the feature-based classifiers (Figure 15).

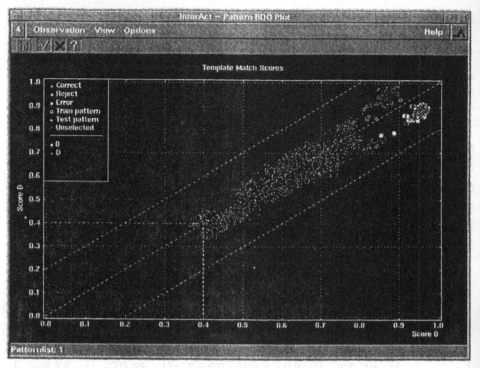

Figure 15. The D/0 class can be separated by a dedicated selection rule on the location with respect to the dashed lines, once this class is already distinguished from all other classes.

In turn, this provides a natural way to provide feature selection information back to the neural nets. Here, thresholds must be set for the voting scheme. By feeding back the classification errors, this measure can be easily adapted.

4.2 Syntax Trees

In the last step of the CLPR system, the results of the character recognition module are compared with some syntactical rules. These rules can be divided into two types. The first set of rules takes into account the spatial requirements of the characters as defined by the official guidelines. For Dutch license plates, the spacing between characters should be zero. The second set of rules detects illegal combinations of numerical and alpha-numerical characters. Legal Dutch license plates contain only pairs of digits and letters, e.g., GD-85-DF, 56-FG-PR or KL-54-68. Furthermore, there are some illegal letter pairs like the SS and SD. A candidate license number should pass both sets of rules in order to be marked as recognized [9]. In case of any conflicts, the image will be rejected (Figure 16).

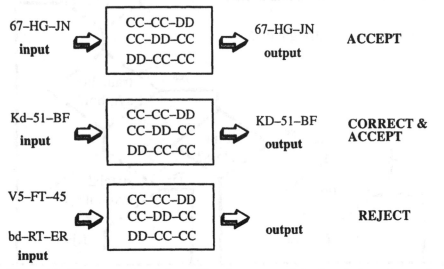

Figure 16. The syntax module verifies the syntax of the found characters and tries to perform a final resolve action on the dubiously recognized characters, shown in small print. Dutch license plates contain three groups each containing either two digits or two characters.

By adding more rules to the syntactical analysis module, it is possible to decrease the error rate at the cost of a higher rejection rate. In [7] the spatial requirements and in [26] the syntactical requirements are used to fill in or replace characters with more likely ones. Experiments showed that this should be done only as a last resort, as the recognition rate increases only marginally at the expense of a slightly higher error rate.

4.3 The Overall Architecture

The VIPUR system (Figure 17) starts by taking out the character "1," if present. This is because the cut-out for this character has such special geometrical features that it stands out from all other characters. The set of all other characters will be split on the basis of the restrictions on the learning performance of the feature-based classifiers. If a symbol is too rare to be sufficiently trainable, the recognition moves directly to template matching; otherwise, features will be derived and tested. If the voted evaluation of the neural networks leaves the decision to be one of the members of a class of characters, a dedicated template matcher will be used.

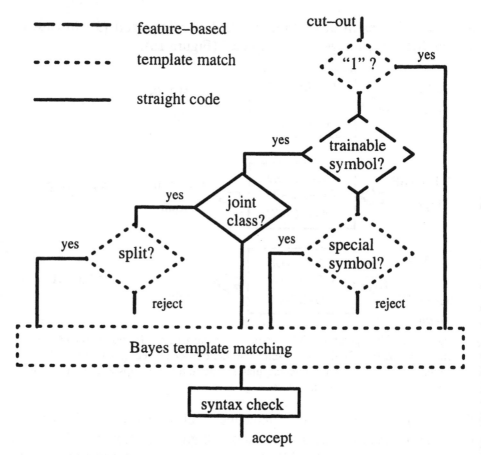

Figure 17. An overall look on the classification strategy.

Even when the cut-out has been classified with a reasonable accuracy, a separate Bayes decision network will be applied to inspect whether or not this decision can be supported from template matching. As a last aid in achieving the highest recognition quality, the set of classified characters will be inspected for conformity with the national license plate regulations.

As a last resort for the rejected cut-outs, we change the cut-out hypotheses. From a template matching for the original cut-out (Figure 18a), the five runner-ups are taken (Figure 18b) and for each of them, a new cut-out is made on the assumption that it may be the correct character. Then again, a feature-based classification is made (Figure 18c). If the classification is in agreement for all five new cut-outs, the character is still assumed to be correctly recognized.

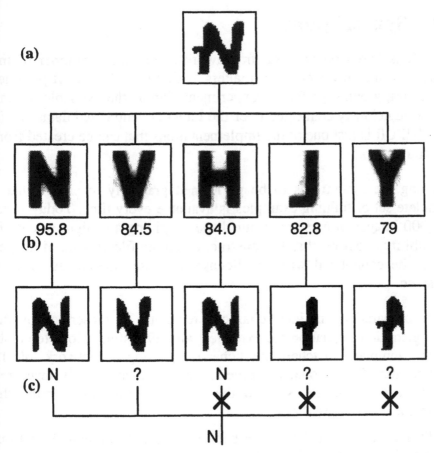

Figure 18. An example of template matched character isolation. The original cut-out (a) is matched to all templates and five runner-ups (b) are selected. On the hypothesis of these characters, new cut-outs are made (c) and classified. Recognition is performed if all classifications are in agreement.

VIPUR Architecture and Development

here is no such a thing as a universal license plate recognition system. he type and quality of the cameras may change, the position and angle f cameras with respect to the driving vehicles are not standardized, cense plates differ per country, and even within a country, the license late may change. In other words, from time to time and from country country, the specification of the license plate recognition system will ary and the system must be constructed such that maintenance is latively easy to perform. This is the basic philosophy that underlies e development of the VIPUR system.

5.1   System Synthesis

VIPUR is based on an open framework containing a number of the classifiers and Bayes decision makers as discussed before. It provides the testing ground for further experimentation as the available tooling can be judiciously connected over the InterAct centralized database. In fact, VIPUR is just one of the implementations that can be created from this framework.

Training the system has to be performed gradually to ensure that a complete set of training examples is available every time. Today some 600,000 images of cars are at our disposal, giving a straight stochastic plausibility to our results. To construct a training file, this set of images has to be connotated with the license plate information as visually retrieved.

1. In preparation, the ideal character templates as prescribed by the registration regulation are extended with 29 disformations to create the template matchers per character, using a network of 12 Pentium-199 personal computers. The result is visually inspected by scatter plots to establish the reference basis for learning the feature-based classifiers.

2. Further, a dedicated tool has been constructed to manually cut-out the license plates from the presented images and to enter the contained number into the computer. The location of the license plate is used later to validate the automated segmentation; the entered number is used for verification of the learning.

3. Finally, the actual VIPUR system is constructed from the explicit information in the template file and the implicit information in the license plate file. The neural classifiers are allowed to leave some difficult classes still undecided, to be resolved later by dedicated template matchers.

Classifiers are in the first instance created as C-level InterAct calls, thus facilitating a neural net model (InterAct is the name of our neural network development environment). Such models can, in turn, be transformed to C-programs to replace the original set of calls. A step by-step VIPUR C-program is created that is eventually ported to the actual camera-based stand-alone system.

This automated style of development provides a high level of maintainability. For each change in camera-type, view angle and plate, a new VIPUR version can be created without manual intervention. At this moment, VIPUR assembly takes 6 hours on a HP-5000 client-server network in full use. The major effort is rather in bringing up a new set of vehicle images to benchmark the practical experiments going on at the road side.

5.2 System Verification

Where the VIPUR development methodology is targeted for quality of engineering and maintenance, the details of the engineering are directed to the flexibility in achieving the right compromise between the various internal handles.

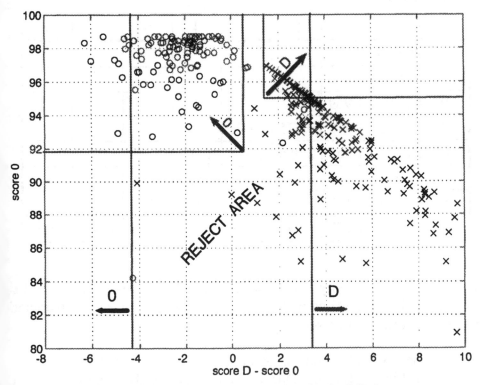

Figure 19. A typical clustering problem for the D/0 class.

The key to proper recognition is the clustering of the templates (Figure 19). A number of techniques are advocated to quantify the quality of a given clustering, such as

- the distance/overlap of the clusters,
- the information content of the clusters,
- the correlation between clusters.

As an example, Table 1 shows the in-class and between-class distance of nine clusters based on digit template matching. A comparison of these two measures provides some insight into the potential difficulties to separate clusters through the use of features. In [10] we have alternatively explored the use of learning conflicts; here we simply evaluate the cluster quality from the resulting ease of classification.

Table 1. The in-class variance and the between mean-class distance of 8 template-based feature sets. The two simple measures give an indication of the possible difficulties to separate classes. The diagonal elements contain the class variance.

	0	1	2	3	4	5	6	7
0	0.005							
1	2.000	0.006						
2	1.885	1.969	0.02					
3	1.803	1.918	0.875	0.021				
4	1.776	1.153	2.056	1.918	0.02			
5	1.893	2.504	1.112	1.115	2.350	0.005		
6	1.387	2.128	1.644	1.557	1.702	1.508	0.009	
7	1.907	1.273	1.496	1.489	1.512	1.828	1.906	0.004

The result can be easily overviewed by graphical means, such as the windmill diagram, showing the scoring obtained for each character from the current feature set (Figure 20).

The ultimate visualization is, of course, the scatter plot. As already shown in Figure 15, the D/0 class can be easily separated by a dedicated template matcher, where the general-purpose matcher (Figure 19) is confronted with severe separation problems.

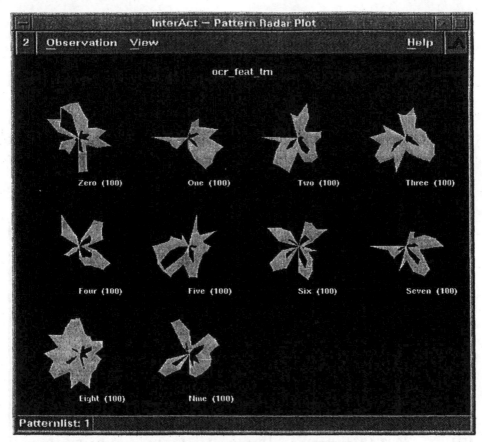

Figure 20. Graphical representation of the template matching scores for all digit templates.

5.3   Evaluation

In most articles the performance of the license plate recognition system is characterized by the recognition rate and error rate as defined by Formula (3) and Formula (4).

$$recognition\ rate = \frac{(number\ of\ correctly\ read\ characters)}{(number\ of\ found\ characters)} \qquad (3)$$

$$error\ rate = \frac{(number\ of\ badly\ read\ characters)}{(number\ of\ found\ characters)} \qquad (4)$$

As the number of false accepts depends on the allowance for false reject, it is necessary to state the rejection rate as defined by Formula (5).

$$rejection\ rate = \frac{(number\ of\ rejected\ characters)}{(number\ of\ found\ characters)} \tag{5}$$

The number of correctly read, badly read and rejected characters should add up to the total number of characters. Regretfully, the rejection rate is seldom mentioned, which makes a comparison between system performances very difficult. Table 2 overviews the literature on car license plate recognition systems for the claimed results on character restructuring.

Table 2. An overview on character level of CLPR systems.

Reference	Recognition/Error Rate	Remarks
[25]	95.5% / 0.9%	Input: 20×30 binary pixels for each character, MLP network. Rejection rate 3.6%.
[2]	98.2% / 1.8%	Segmentation and recognition with MLP networks with pixel input.
[28]	95% / 5%	Characters between 16×23 and 38×47 pixels. MLP network with features input.
[7]	98.2 % / ? (digits) 96.1% / ? (Chin. chars.)	Input: 512×480 gray level images. Character alignment to fill obscured digits.
[8]	96.8% / 3.2%	Classification based on neural network with 12 input features. No details.
[1]	90% / ? (daytime) 65% / ? (nights)	Rates are based on license plates. Based on template matching.
[5]	97% / 1%	768×493 gray level images. Segmentation with fuzzy logic. Classification with MLP network with pixel input (14×14) with additional top and bottom pixel inputs (14×9).
[29]	94% / ?	Classification with 12×16 pixel input MLP.
this approach	99.1% / 0%	Classification with multiple neural networks with feature inputs and/or template matching. Rejection rate 0.9%.

Our CLPR system has been tested on almost 600,000 different images. For the readable plates and isolated characters, a recognition rate of 99.1% and a rejection rate of 0.9% per character is obtained, while an error rate could not be determined because of the lack of false readings.

This gives a rosy view of the system performance, but a more realistic measure should be based on the recognition of the real images rather than on single, isolated characters per presegmented plate. For this purpose we have to revise the formulae for recognition, error and rejection rate to

$$recognition\ rate = \frac{(number\ of\ correctly\ read\ plates)}{(number\ of\ found\ plates)} \tag{6}$$

$$error\ rate = \frac{(number\ of\ badly\ read\ plates)}{(number\ of\ found\ plates)} \tag{7}$$

$$rejection\ rate = \frac{(number\ of\ rejected\ plates)}{(number\ of\ found\ plates)} \tag{8}$$

Table 3. A comparison between complete CLPR systems

Reference	Recognition/Error Rate	Remarks
[3]	61.7% / 11.2% (low speed) 33.6% / 6.6% (high speed)	Evaluation of the 3M RIA-300 CLPR system. Rates are based on license plates. Rejection rate of 27.1% at low car speeds and 59.8% at high car speeds.
[26]	63% / 37%	Results produced with syntax forcing. Rates are based on license plates.
[27]	76% / ?	Rate base on license plates (7 characters). Input: 512x512 gray level image. Classification based on features.
this approach	85.6% / 0.001%	Segmentation with DT-CNNs. Classification with multiple neural networks with feature inputs and/or template matching. Rejection rate 14.4%.

To our sincerest regret, we have not obtained more recent performance figures on any commercial CLPR system. To set a target, the Dutch government has stated that a recognition rate better than 70% and a error rate less than 0.02% would be acceptable. This seems, in the first instance, a harsh ruling; on the other hand, our experience has shown that such performance is in reach when using multiple experts. Recent experience has indicated that further improvements can be achieved by a proper balance between error and rejection rate.

6 Summary

The VIPUR system has come a long way since the first (naive in retrospection) attempts in 1994. In the course of time, our understanding of the intrinsic problems has matured and our need for a proper and maintainable working environment has risen. Where the first version was constructed for a singular product, we have eventually arrived at a flexible toolbox that could also produce a license plate recognition system. In the meantime, the ease of turning the tools for the problem at hand has already been demonstrated in a number of other projects ranging from document retrieval to sensory diagnosis and control.

Compared to the first attempts, the major difference is in the introduction of mixed experts to achieve an improved overall recognition rate. Originally we had to rely on general-purpose template matching to achieve a reasonable error rate. The improvements in feature-based classification has changed its role to a mere point of reference. In return, we have introduced a number of small but dedicated template matchers for final decision making. This architecture allows making many, easy decisions as fast as possible, thereby creating room for a more comprehensive analysis of the few complicated problems.

Another important characteristic is the "delayed binding" of disputable decisions. Using the hierarchical expert structure, decisions are postponed until maximum confidence is obtained. Eventually even new cut-outs are created to improve recognition before a character reading is finally rejected.

In the laboratory, a 91% recognition rate has been achieved, but this still needs real life experimental verification on the road. Current realism is an 85% recognition rate. The improvement in flexibility has become noteworthy, where the end user feels the political need to be able to tune the system performance along the road. The trade off between true and false acceptances is no longer buried within the system, but has become a handle with an external grip. Such a wish can be easily accommodated.

References

[1] Kanayama, K., Fujikawa, Y., Fujimoto, K., and Horino, M. (1991), "Development of Vehicle-License Number Recognition System using Real-Time Image Processing and its Application to Travel-Time Measurement," *Proceedings of the 41st IEEE Vehicular Technology Conference* (St. Louis, MO), pp. 798-804.

[2] Föhr, R. and Raus, M. (1994), "Automatisches Lesen amtlicher Kfz-Kennzeichen," *Elektronik*, (in German), No. 1, pp. 60-64.

[3] Davies, P., Emmott, N., and Ayland, N. (1990), "License Plate' Recognition Technology for Toll Violation Enforcement," in: *Proceedings of IEE Colloquium on Image analysis for Transport Applications*, Vol. 35, pp. 7/1-7/5.

[4] Robertson, D.J. (1993), "Automatic Number Plate Reading for Transport Applications," *Proceedings of IEE Colloquium on Electronics in Managing the Demand for Road Capacity*, Digest No. 1993/205, pp. 13/1-13/4.

[5] Hwang, C., Shu, S., Chen, W., Chen, Y., and Wen, K. (1992), "A PC-Based License Plate Reader," *Proceedings Machine Vision Applications, Architectures and Systems Integration*, Vol. SPIE-1823 (Boston, MA), pp. 272-283.

[6] Mori, S., Suen, C.Y., and Yamamoto, K. (1992), "Historical Review of OCR Research and Development," *Proceedings of the IEEE*, Vol. 80, No. 7, pp. 1029-1057.

[7] Tanabe, K., Marubayashi, E., Kawashima, H., Nakanishi, T., and Shio (1994), A., "PC-Based Car License Plate Reading," *Image and Video Processing*, Vol. SPIE-2182, pp. 220-231.

[8] Abe, S., and Lan, M. (1993), "A Classifier Using Fuzzy Rules Extracted Directly from Numerical Data," *Proceedings of the Second IEEE International Conference on Fuzzy Systems*, Vol. 2 (San Francisco, CA), pp. 1191-1198.

[9] Nijhuis, J.A.G., terBrugge, M.H., Hettema, H., and Spaanenburg, L. (1995), "License Plate Recognition by Fuzzy Logic," *Proceedings of the 5th Aachener Fuzzy Symposium EUFIT'95* (Aachen, Germany), pp.147-153.

[10] Nijhuis, J.A.G., terBrugge, M.H., Helmholt, K.A., Pluim, J.P.W., Spaanenburg, L., Venema, R.S., and Westenberg, M.A. (1995), "Car License Plate Recognition with Neural Networks and Fuzzy Logic," *Proceedings of the ICNN'95*, Vol. V (Perth, Western Australia), pp. 2232-2236.

[11] terBrugge, M.H., Nijhuis, J.A.G., and Spaanenburg, L. (1998), "License Plate Recognition using DT-CNNs," *Proceedings International Workshop on Cellular Neural Networks CNNA'98* (London, U.K.), pp. 212-217.

[12] Harrer, H., and Nossek, J.A. (1992), "Discrete-Time Cellular Neural Networks," *International Journal of Circuit Theory and Applications*, Vol. 20, pp. 435-467.

[13] Agui, T., Choi, H.J., and Nakajima, M. (1988), "Method of Extracting Car Number Plates by Image Processing," *Systems and Computers in Japan*, Vol. 19, No. 3, pp. 46-52.

[14] Postolache, A., and Trecat, J. (1995), "License Plate Segmentation Based on a Coarse Texture Approach for Traffic Monitoring System," *Proceedings of the ProRISC & IEEE-Benelux Workshop on Circuits, Systems and Signal Processing* (Mierlo, The Netherlands), pp. 243-250.

[15] terBrugge, M.H., Krol, R., Nijhuis, J.A.G., and Spaanenburg, L. (1996), "Design of Discrete-Time Cellular Neural Networks Based on Mathematical Morphology," *Proceedings International Workshop on Cellular Neural Networks CNNA'96* (Sevilla, Spain), pp. 1-5.

[16] terBrugge, M.H., Stevens, J.H., Nijhuis, J.A.G., and Spaanenburg, L. (1998), "Efficient DT-CNN Implementations for Large-neighborhood Functions," *Proceedings International Workshop on Cellular Neural Networks CNNA-98* (London, U.K.), pp. 88-93.

[17] terBrugge, M.H., Spaanenburg, L., Jansen, W.J., and Nijhuis, J.A.G. (1996), "Optimizing the Morphological Design of Discrete-Time Cellular Neural Networks," *Proceedings International Workshop on Cellular Neural Networks CNNA'96* (Sevilla, Spain), pp. 339-343.

[18] terBrugge, M.H., Nijhuis, J.A.G., and Spaanenburg, L. (1998), "Transformational DT-CNN Design From Morphological Specifications," *IEEE Transactions on Circuits and Systems-I: Fundamental Theory and Applications*, Vol. 45, No. 9, pp. 879-888.

[19] Trier, O.D. and Taxt, T. (1995), "Evaluation of binarization methods for document images," *IEEE Transactions on Pattern Analysis and Machine Intelligence*, Vol. 17, Nr. 3, pp. 312-315.

[20] Hu, M.-K. (1962), "Visual pattern recognition by moment invariants," *IRE Transactions on Information Theory*, Vol. 8, pp. 179-187.

[21] Trier, O.D., Jain, A.K., and Taxt, T. (1996), "Feature extraction methods for character recognition," *Pattern Recognition*, Vol. 29, Nr. 4, pp. 641-662.

[22] Pratt, W.K. (1991), *Digital Image Processing*, 2nd ed., John Wiley & Sons, New York.

[23] Haykin, S. (1994), *Neural networks: a comprehensive foundation*, MacMillan, New York.

[24] Glauberman, M.H. (1956), "Character recognition for business machines," *Electronics*, Vol. 29, pp. 132-136.

[25] Lisa, F., Carrabina, J., Pérez-Vincente, C., Avellana, N., and Valderrama, E. (1993), "Two-bit Weights Are Enough To Solve Vehicle License Plate Recognition Problem," *Proceedings of the International Conference on Neural Networks*, Vol. 3 (San Francisco, CA), pp. 1242-1246.

[26] Williams, P., Kirby, H., Montgomery, F., and Boyle, R. (1989), "Evaluation of Video-Recognition Equipment for Number-Plate Matching," *Proceedings of the 2nd International Conference on Road Traffic Monitoring* (London, U.K.), pp. 89-93.

[27] Lotufo, R.A., Morgan, A.D., and Johnson, A.S. (1994), "Automatic Number-Plate Recognition," *Proceedings of the IEE Colloquium on Image Analysis for Transport Applications*, Vol. 35, pp. 6/1-6/6.

[28] Yoo, J., Chun, B., and Shin, D. (1994), "A Neural Network for Recognizing Characters Extracted from Moving Vehicles," *Proceedings of World Congress on Neural Networks*, Vol. 3 (San Diego, CA), pp. 162-166.

[29] Kertész, A., Kertész, V., and Müller, T. (1994), "An On-line Image Processing System for Registration Number Identification," *Proceedings of IEEE International Conference on Neural Networks*, Vol. 6 (Orlando, FL), pp. 4145-4147.

INDEX

Milton Keynes UK
Ingram Content Group UK Ltd.
UKHW021618071024
449327UK00020BA/1102